KU-662-801

Garden Pests
and Diseases

Garden Pests and Diseases

Dr. Joe Stubbs

WARD LOCK LIMITED · LONDON

© Ward Lock Limited 1979

First published in Great Britain in 1979
by Ward Lock Limited, 116 Baker Street,
London W1M 2BB, a Pentos Company.

All Rights Reserved. No part of this publication
may be reproduced, stored in a retrieval system,
or transmitted, in any form or by any means,
electronic, mechanical, photocopying,
recording, or otherwise, without the prior
permission of the Copyright owners.

House editor Frances Dixon

Text filmset in Times
Set, printed and bound in Great Britain by
Cox & Wyman Ltd,
London, Fakenham and Reading

British Library Cataloguing in Publication Data

Stubbs, Joe
    Garden pests and diseases. – (Concorde
books).
    1. Garden pests   2. Plant diseases
    I. Title   II. Series
    635'.04'9   SB603.5

    ISBN 0–7063–5365–X
    ISBN 0–7063–5367–6 Pbk

Contents

ACKNOWLEDGEMENTS

The publishers gratefully acknowledge the following persons, agencies and companies in providing and granting permission to reproduce the following photographs: Dr. A. Beaumont (p. 104); Dr. J. Stubbs (pp. 69, 79 and 103 (below)); Heather Angel (p. 36); A. G. Heath/Biofotos (pp. 85 and 86); Harry Smith Horticultural Photographic Collection (pp. 35 and 103 (top)); Imperial Chemical Industries Plant Protection Ltd. (pp. 18, 25, 70, 75, 81 and 113); and Murphy Chemical Ltd. (p. 17).

1 Identification and Control

Cultivated plants are under constant threat of attack by a whole host of enemies. Larger pests such as deer, rabbits, squirrels, dogs, cats and birds are easy to identify and can be guarded against by netting, scarers and repellents. Quite a different problem is posed by the smaller pests which range from slugs, insects and mites to the microscopic eelworms. With these, the first difficulty is that of identification since they come in such a wide variety of shapes and sizes. Admittedly, some pests such as greenfly, blackfly and caterpillars are easily spotted and readily recognized. Other pests, which feed at night, are not normally seen on the plants during the daytime. Root-feeding pests are equally difficult to detect. In these situations it is therefore necessary to be able to identify pests on the basis of the damage they cause. Then, having successfully identified the cause of the trouble, it is helpful to know something about the life cycle of the pest in order to design appropriate counter-measures.

Identification of plant diseases also presents difficulties because these organisms are microscopic in size. Here again, the practical approach is to use the plant symptoms as a basis for recognition.

NATURAL CONTROL

In a natural environment, predators and parasites have a major regulating effect on the pest population. For instance, ladybirds and their larvae eat large quantities of aphids. The larvae of lacewings and hover flies also feed on these pests. Red spider mites on the other hand are controlled by the black-kneed capsid bug whilst caterpillars may be parasitized by the larvae of various small parasitic wasps. There are also indications that ground beetles feed on the eggs of the cabbage root fly.

This system of natural pest control produces a satisfactory balance of plants, pests and predators in the wild. Unfortunately, however, it does not normally give an acceptable level of pest control in the garden. Some success, however, has been achieved in commercial greenhouses by the introduction of Encarsia formosa, which parasitizes whiteflies, and by the use of predatory mites for the control of red spiders. This approach is unfortunately not generally feasible in the amateur greenhouse because these predators and parasites require fairly high temperatures if they are to operate successfully.

CHEMICAL CONTROL
The use of modern garden pesticides provides a relatively easy and effective method for the control of pests and diseases. These products when stored, handled and applied as directed, are not only highly effective but are safe to the user, his family, pets and wildlife. In this connection it must be emphasized that garden chemicals are subject to intensive testing over a period of years to ensure that they are not only highly active against pests or diseases but are also safe to use and have minimal effects on the environment. The results of this research are then passed to a committee of independent scientists for clearance under the Government's Pesticides Safety Precautions Scheme and it is only after such clearance has been granted that the new chemical is marketed.

One key to the successful use of garden pesticides is to follow the directions for use which are given on the product label. Increasing the dosage, for instance, will not improve the pesticidal activity and may well result in damage being caused to the treated plants. Should, however, both pest and disease control be needed at the same time, then it is possible to mix an insecticide with a fungicide to produce a combined spray. Attention should also be paid to any warning on the product label that the material is not suitable for use on certain plants or at particular growth stages. Incidentally, plants which are particularly prone to chemical damage are cacti, succulents, ferns and orchids.

Plant damage can also occur if the sprays are applied in bright sunshine because the droplets can act as lenses, concentrating the sunlight and producing burn marks on the plant tissue. It is therefore advisable to delay spray treatments until early evening. This timing has the additional advantage that bees will have stopped working among the flowers and so will not be at risk from insecticides.

NAMING OF PESTICIDES

Most modern pesticides are extremely complicated chemicals with long and involved names. For this reason they have been given shorter common chemical names. These are listed in the Appendix on page 115, together with a selection of the proprietary product names under which they can be bought. The active contents of other garden pesticides which have not been included in this short list can be found on the product labels.

SELECTION OF SUITABLE PESTICIDES

Most insecticides are active against a range of plant pests. There are, however, exceptions to this general rule. Pirimicarb, for instance, is mainly active only against greenfly and blackfly. It can therefore be safely used against these pests without risk to their natural predators such as lacewings, ladybirds and hover flies. Similarly, bromophos, chlorpyrifos and diazinon are mainly used against soil pests, whilst chlordane is of special interest in the control of earthworms and ants. Then, too, there are some insecticides which are of special value in the control of certain difficult pests such as whiteflies, red spider mites and scale insects.

Most fungicides are also active against a range of diseases. Some, however, are only active against powdery mildews whilst others are of no value against this type of disease.

In view of the varied response of pests and diseases to different chemicals it is desirable first of all to identify the cause of the trouble and then to select and apply the appropriate remedy. This, therefore, is the approach used in the succeeding chapters.

TIMING OF TREATMENTS

In general, the application of insecticides can be deferred until the first signs of attack. This, however, is too late to be fully effective against some soil pests. Consequently, protective treatments should be applied at, or shortly after, seeding or planting out. This also applies to fungicides used to control some soil-borne diseases. Fungicides used against foliar diseases can be applied at the first sign of damage but it must be accepted that established infections will not be eliminated. It is therefore better practice to carry out a programme of protective sprays against diseases such as apple and pear scab, apple mildew, and rose blackspot and mildew. Insecticidal sprays can then be included in this programme as necessary.

THE SAFE HANDLING OF PESTICIDES

Since the undiluted products may be poisonous if eaten or drunk it is most important that pesticidal products are stored safely away from children and pets. The diluted sprays present much less risk but, even so, cats and dogs must not be given the opportunity of drinking the spray solution from buckets or other open containers. It must, however, be emphasized that the dried residues on sprayed plants present no hazards to pets.

When preparing the spray solution care should be taken to avoid getting drops of the concentrate on exposed skin. Immediate washing of the affected part, however, removes the risk of harmful effects. It is also advisable to wash both hands and face after completing the spray application in order to remove any possible contamination by chemicals.

Amongst the various dust and granular products, slug pellets based on metaldehyde present special problems because they are attractive to dogs and can be poisonous if consumed in quantity. Opened packets of pellets, therefore, need to be kept out of the reach of pets. It is also advisable to apply only a light scattering of pellets on the ground. Incidentally, this type of distribution is just as effective for the control of slugs and snails as excessively heavy applications.

Spray residues on treated plants quickly break down into harmless materials and so present no hazard to consumers of sprayed crops provided that the recommended interval between the last application and harvesting is observed. Information on this safe period is given on the product label.

Spray-strength pesticides in general lose their activity very quickly. Consequently, surplus spray solution must not be stored for future use. Any excess solution can be safely disposed of by pouring it down the outside kitchen drain. Alternatively, it can be sprayed on the ground under the treated plants provided that it is not allowed to form puddles from which pets could drink. Finally, the sprayer should be thoroughly washed out with water before being put away for future use.

2 Vegetables

ARTICHOKE, GLOBE
Root aphid These greyish-brown coloured aphids, which are covered with a white waxy powder, feed on the plant roots. Control by applying a heavy soil drench of diazinon or pirimiphos-methyl.

Petal blight This shows as brown circular patches on the flower heads. These spots later coalesce and cause rotting of the heads. Protection against this disease is given by fortnightly sprays with zineb.

ARTICHOKE, JERUSALEM
Sclerotinia disease This attacks the stems, causing them to rot above soil level. It can also cause a storage rot of the tubers. Lift and destroy any plant showing stem damage and store only sound tubers.

ASPARAGUS
Asparagus beetle These small beetles have square yellow markings on the black wing cases whilst the front of the body is dull red. They begin their attack in May, eating the foliage and laying eggs on the plants. The grey-black grubs also feed on the foliage and stems. Spray or dust with a general insecticide at the first signs of attack.

Slugs These can damage the spears and should be controlled by scattering slug pellets around the plants.

12

Violet root rot This causes the top growth to turn yellow and die. Affected roots have a purplish covering of fungal threads. It is soil-borne so diseased patches should be isolated by burying a 30cm (12in) band of polythene film round the infected area.

BEANS, ALL TYPES
Bean seed fly The white maggots live in the soil and feed on both the seeds and the young seedlings. It is most serious in cold wet soils, so good soil preparation reduces the intensity of attack. Control by dusting the open drills with bromophos, HCH or pirimiphos-methyl.

Blackfly These are found both on the stems and on the underside of the leaves. They are readily controlled by special greenfly sprays or with general insecticides.

Red spider mite These tiny mites, which are only just visible to the naked eye, feed on the underside of the leaves, causing yellow speckling of the upper leaf surface. Usually most troublesome in hot weather when they can cause the death of the plant. Control by repeat sprays of dimethoate, formothion or pirimiphos-methyl.

Foot and root rots These soil-borne fungi build up if leguminous crops are grown continuously on the same site. Crop rotation is therefore important. A seed dressing containing captan should also be used. Some check to the disease is given by watering the soil with captan, Cheshunt compound or zineb.

BEANS, BROAD
Pea and bean weevil These pests bite U-shaped notches in the leaf margins but the damage is not generally serious. Control by dusting the plants and the surrounding soil with HCH.

Bean rust This disease shows as numerous brown powdery spots on the underside of the leaves. Although it is not very common it can be troublesome in some years. Control by repeat sprays of copper, maneb or thiram.

Chocolate spot This disease first shows as small chocolate-coloured spots on the foliage and stems. In severe attacks the spots run together causing large blackened areas. Actively growing plants, particularly if they are protected from black-fly, are least susceptible. Protection against this disease is given by regular sprays with a copper fungicide.

BEANS, DWARF, FRENCH AND RUNNER
Halo blight This seed-borne disease causes angular spots on the leaves. Each spot is surrounded by a light-coloured halo. In wet weather, bacteria ooze from the spots and so spread the disease. It is prevented by buying good quality seed.

BEETROOT
Mangold fly The grubs, which hatch from eggs laid on the underside of the leaves, burrow into the leaves producing blotch mines. Light attacks can be dealt with by picking off the infested leaves. More widespread infestations can be controlled by spraying with dimethoate, formothion, pirimiphos-methyl or trichlorphon.

Damping-off Seedlings collapse at ground level due to attack by various soil-borne fungi. The best way of avoiding this trouble is to sow thinly and to avoid overwatering.

Leaf spots Various leaf spot diseases can develop but these are usually only serious in wet weather. Prevent further spread by removing infected leaves.

Heart rot This shows first as a browning of the inner tissues of the root. This is followed later by blackening of the root and the death of most of the leaves. It is caused by a deficiency of available boron in the soil and the curative treatment is to apply borax at the rate of 34g per 20sq m (1oz per 20sq yd).

BRASSICAS
Cabbage root fly Eggs are laid on the soil and these hatch into whitish maggots which burrow down and feed on the plant roots. This results in a slowing down of growth and there is a tendency for the plants to wilt in warm weather. Another sign

of attack is that the leaves develop an unhealthy blue colour. Seed rows and young transplants should be protected by applying dusts or granules containing bromophos, chlorpyrifos, diazinon or pirimiphos-methyl. Late attacks on established plants can be dealt with by applying heavy soil drenches of spray-strength pirimiphos-methyl or trichlorphon.

Cabbage mealy aphid This grey-coloured aphid is particularly damaging to young plants where it can kill the growing points, preventing further growth. On established plants it forms dense colonies on the underside of the leaves, and it can also infest Brussels sprouts and broccoli heads. Control by greenfly killers or general insecticides.

Cabbage whitefly These tiny white moth-like insects can be very troublesome. Both the adults and the greenish scale-like larvae feed on the underside of the leaves, sucking up the sap and producing quantities of syrupy honeydew on which unsightly sooty moulds develop. Whitefly are difficult to control because the eggs, and some of the larval stages, are resistant to most insecticides. Best results are obtained with a programme of three or more sprays of bioresmethrin, pirimiphos-methyl or resmethrin, applied every 3–4 days.

Caterpillars The caterpillars of the large cabbage white butterfly feed on the outer leaves. However, more serious damage is caused by the smaller caterpillars of the small cabbage white butterfly and the cabbage moth because these burrow into the heart of the cabbage. The best safeguard is to inspect the plants regularly and to spray with a general insecticide at the first signs of attack.

Flea beetle These tiny dark-coloured beetles attack seedlings, eating small circular holes in the leaves. They are readily controlled by the use of insecticidal dusts applied to the plants and to the surrounding soil.

Turnip gall weevil *See* Turnip.

Slugs Leaves and stems can be seriously damaged by these pests in wet weather. Control by the use of slug pellets.

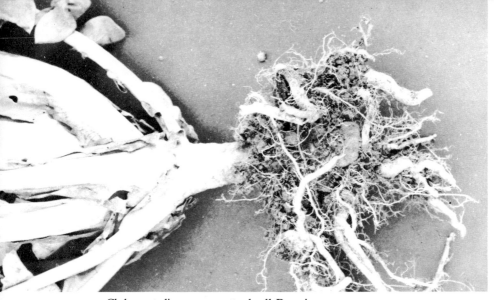
Club root disease can attack all Brassicas

Club root This soil-borne disease causes the roots to thicken and become distorted. Early attacks result in very stunted growth whilst later ones cause discoloration of the leaves and wilting in warm weather. Conditions favouring the disease are poor drainage and acid soil. Ground in which brassicas are to be grown should be improved by the addition of bulky organic matter and by the application of lime. Crop rotation, coupled with weed control, should be practised to reduce the carry-over of the disease. These measures need to be supplemented by the application of calomel (mercurous chloride) dust to the open seed drills, seedling roots and planting holes.

Downy mildew This appears as a white mealy growth on the underside of the leaves. It is most damaging to seedlings, causing a severe check to growth. The best insurance against this disease is to sow thinly, but if it does appear it should be controlled by spraying with dithane or with a copper fungicide.

Leaf spots Spotting of older leaves can be caused by a number of fungi, particularly in wet seasons and on rather soft grown plants. The best immediate action is to remove and burn

Whitefly

Mealybug

Red spider mite
stringing on
French bean

infected leaves. Crop rotation should also be practised to reduce the risk of attack.

White blister Glistening white spots appear on the leaves. These then spread to give a white covering to the leaf surface. Remove and burn infected leaves. Here again, crop rotation helps to reduce the incidence of the disease.

Wirestem The stems of affected seedlings become black and shrunken at soil level. Seedlings attacked by this soil-borne disease often die and those which survive produce only stunted growth. Control by crop rotation, good cultivation and thin sowing.

Brown curds Cauliflowers are sensitive to boron deficiency which causes distortion of the young leaves and leads to the development of small, bitter tasting heads with brown curds. Repeated spraying with a foliar feed containing boron can help. Alternatively, where the trouble is known to exist, apply borax to the soil at 34g per 20sq m (1oz per 20sq yd) before planting.

Whiptail Where there is a deficiency of molybdenum in the soil, the leaves of broccoli and cauliflowers become thin and strap-like. This condition can be remedied by applying a solution of sodium molybdate to the soil at the rate of 28g per 8sq m in 9 litres of water (1oz per 10sq yd in 2gal).

CARROT
Carrot root fly The white maggots of this pest eat the fine feeding roots and later tunnel into the tap root. The first signs of attack are reddening of the leaves and wilting of the foliage in sunlight and this is usually followed by the leaves turning yellow. There are two generations a year. The first generation attacks crops in May and June whilst the second appears in August and September. So, carrots sown in late May and harvested early are generally free from attack. Thin sowing and no thinning reduce the risk of attack. Nevertheless it is recommended that bromophos, chlorpyrifos, diazinon or pirimiphos-methyl should be applied at sowing.

Carrot-willow aphid This aphid not only weakens the plants by feeding on them but also transmits virus disease. It can cause major yield losses, but is readily controlled by greenfly killers or general insecticides.

Black rot Shows as black sunken lesions on mature roots and can cause storage losses. Store only healthy roots and destroy any found to be infected.

Sclerotinia rot This storage disease is indicated by the presence of a white fluffy fungal growth on the surface of the roots. Inspect roots before storage and make regular checks on the stored carrots.

Violet root rot This can be detected on lifting the crop by the presence of a purple surface felting on the roots. Newly harvested roots can be eaten but should not be stored.

CELERIAC *See* Celery.

CELERY
Carrot root fly Maggots of this pest can damage the seedling roots of celery. Control by the addition of diazinon dust to the seedling compost, coupled with repeat treatment at planting out.

Celery fly Larvae of this fly tunnel into the leaf tissue producing large blisters. Control by spraying with dimethoate, malathion or trichlorphon.

Slugs These can be a problem after earthing up. Control by applying slug pellets before earthing up or by watering with a liquid metaldehyde product.

Leaf spot This first appears as brown spots on the leaves and petioles. Later, pinhead fruiting bodies develop on these spots. This disease is not generally troublesome, but if it appears the plants should be given repeat sprays with benomyl, copper or zineb.

Heart rot This bacterial disease, which causes a slimy heart rot, is not common since it only affects damaged tissue. The best protection is to practise crop rotation and to make every effort to control slugs.

CHICORY
Cutworms These are soil-living caterpillars which gnaw the base of the stem. Control by applying a soil insecticide dust before seeding. Later attacks can be dealt with by the use of soil drenches of spray-strength general insecticide.

COURGETTE *See* Marrow.

CUCUMBER
Aphids These not only reduce plant vigour but also spread cucumber mosaic virus. Spray with a greenfly killer or general insecticide at the first sign of attack.

French-fly This glistening, whitish, slow-moving, bristly mite is often present in unrotted manure. It feeds on the leaves and shoot tips of the young plants causing distorted growth or even killing the apical bud. Control by spraying the plants with dimethoate or malathion.

Grey mould This disease, which is commonest in wet seasons, shows as a brownish-grey fungal growth on rotting stems, fruits and leaves. Remove affected parts at the first sign of attack and then apply protective sprays of benomyl or thiram.

Powdery mildew The leaves and stems become covered with a white powdery fungal growth and eventually die. Control by repeat sprays of benomyl, dinocap, sulphur or thiophanate-methyl.

Stem and root rots Various organisms can cause this type of disease and these attacks are favoured by under- or over-watering and by repeated use of the same growing site. Reduce the risk of attack by growing the crop on a fresh site each year and by careful watering. Should it develop, the disease can be checked by watering the soil with captan, Cheshunt compound or zineb.

Cucumber mosaic virus This is a common and serious problem. Infected plants are readily recognized by their puckered, mottled leaves and by their stunted growth. Such plants should be destroyed and the spread of the disease checked by effective control of aphids which transmit the disease.

ENDIVE
Generally trouble-free.

GARLIC
White rot This soil-borne disease affects the roots and the base of the bulb where it shows as a white fluffy growth. Mainly controlled by crop rotation. The application of calomel (mercurous chloride) dust at planting is also helpful.

LEEK
Pests, general *See* Onion pests as these also attack leek.

Rust This disease shows as elongated clumps of bright orange spores on the leaves. It is favoured by a high nitrogen content in the soil caused by over-feeding. Difficult to control but the infected plants are still edible.

White rot *See* Onion.

LETTUCE
Aphids Greenfly weaken the plants by feeding on the sap and also transmit virus disease. Easily controlled by the use of greenfly killers or general insecticides. Root aphids, which have a white powdery covering, feed on the roots causing stunting of the growth and yellowing of the leaves. Control by the application of soil drenches of spray-strength HCH, malathion or pirimiphos-methyl.

Cutworms These soil-dwelling caterpillars gnaw the base of the stem and cause the plants to collapse. They can be controlled by the use of a soil insecticide at sowing. Later attacks can be dealt with by the use of soil drenches of spray-strength general insecticides.

Slugs These pests are liable to attack lettuce at all stages of growth. Control by the use of slug pellets.

Damping-off Attacks by various soil-borne organisms can cause the seedlings to collapse at ground level. The use of a captan seed dressing reduces the risk of attack but the best method of control is to sow thinly and to avoid overwatering.

Downy mildew This disease causes yellow patches on the upper surface of the leaves with corresponding mealy patches on the underside. Affected leaves later turn brown. The risk of attack is reduced by not overcrowding the plants and by spraying with benomyl, captan, thiram or zineb.

Grey mould This disease, which shows as a grey fungal growth, usually attacks near the base of the plant and can lead to wilting. Affected leaves should be removed and the plants sprayed with benomyl or thiram.

MARROW
Aphids Greenfly not only weaken the plant but also transmit cucumber mosaic virus. Spray at the first sign of attack, using either a greenfly killer or a general insecticide.

Slugs These damage both the foliage and the fruit. Control by spreading slug pellets around the plants.

Foot and root rots These are caused by various soil organisms and the attacks are favoured by both under- and overwatering. There is no fully effective method of treatment, but watering with Cheshunt compound or zineb may check the disease on slightly infected plants.

Grey mould This is not a common disease except in wet seasons when a brownish-grey fungal covering can develop on rotting stems, fruits and leaves. Remove affected parts and then spray the plants with benomyl or thiram.

Powdery mildew Shows as a powdery white covering on the leaves and stems. Control by repeat spraying with benomyl, dinocap, sulphur or thiophanate-methyl.

Cucumber mosaic virus *See* Cucumber.

MELON *See* Cucumber.

ONION
Onion fly The maggots of this fly attack the roots and bulb, the first sign of attack being that the outer leaves turn yellow and wilt. Protection against this pest is given by the application of bromophos, chlorpyrifos or diazinon granules at planting. Alternatively, HCH or pirimiphos-methyl dusts can be applied when the seedlings are at the loop stage with a second treatment given 3 weeks later.

Thrips In hot weather these tiny slender insects can cause a white flecking of the foliage. Readily controlled by most general insecticides.

Onion eelworm These microscopic soil-borne pests enter the plant tissue causing both the leaves and the bulbs to become swollen and distorted. There is no available chemical cure and the only answer is to avoid growing onions on infected soil for several years.

Neck rot A fungus disease which can cause large losses in storage. It first shows as a mouldy growth around the neck of the bulb and later causes a general soft rot. Ensure that the bulbs are well ripened and that they are stored in dry, well-ventilated conditions.

Onion downy mildew Infected leaves turn greyish before wilting and falling over. In moist conditions, when the disease is most serious, the leaves may show a purplish bloom. The main remedy is to practise crop rotation, but spraying with zineb also helps to reduce the spread of the disease.

White rot The first sign of attack by this soil-borne disease is

24

yellowing and wilting of the leaves brought about by the damage to the roots. The fungus then spreads round the bulb as a white fluffy growth. Control by crop rotation supplemented by the application of calomel (mercurous chloride) dust to the open seed drills.

PARSNIP

Carrot fly The small maggots of this pest damage the roots. For control measures see 'Carrot'.

Celery fly Larvae of this fly tunnel into the leaves, causing large blisters. Control by spraying with dimethoate, malathion, pirimiphos-methyl or trichlorphon.

Parsnip canker This very common disease causes rotting of the top of the root during the late autumn and winter. Crop rotation reduces the incidence of the disease but the best insurance is to grow canker resistant varieties such as 'Avonresister'.

PEA

Pea and bean weevil These small beetles eat U-shaped notches in the leaf margins but the damage is not generally serious. They can be controlled by sprinkling insecticidal dust on to the plants and surrounding soil.

Pea and bean weevil on broad bean

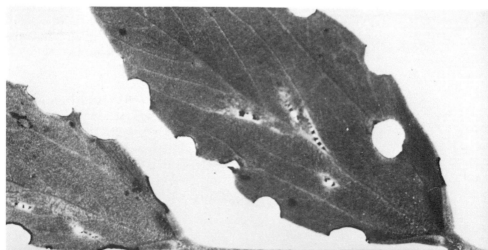

Pea moth Peas flowering over the period from early June to mid-August are liable to attack by this pest. The moth grubs tunnel into the young pods and eat the developing seeds. Damage can be prevented by spraying with fenitrothion or pirimiphos-methyl about a week after the plants start to flower.

Pea thrips These small, sap-feeding insects cause a silvery discoloration of the leaves and shrivelling of the pods. At the first sign of attack the plants should be sprayed with a general insecticide.

Foot roots and wilts Affected plants yellow and wilt and their cropping is reduced. The continuous growing of peas on the same site leads to a build-up of soil organisms causing these troubles. Crop rotation is therefore the best insurance and this should be supplemented by the use of a captan seed dressing.

POTATO
Wireworm This is mainly a pest of newly cultivated ground. The thin orange-coloured larvae, which can be up to 2.5cm (1in) long, tunnel into the tubers. If this pest is known to be present, the soil should be treated with bromophos, chlor-pyrifos or diazinon granules at planting time.

Slugs Slug damage to potatoes is caused by underground-dwelling, black keeled slugs which begin their attack in August. Tubers can be almost completely hollowed out by the slugs. There is no completely effective method of chemical control so the best solution is to restrict cropping to early varieties and to lift the crop as soon as possible.

Potato cyst eelworm This microscopic pest, which is all too common in some areas, can cause serious loss of crop. Its presence can be detected by the development of pin-head sized cysts on the plant roots. These are white at first but later turn orange-brown. Infected plants die early and produce only tiny tubers. In the absence of effective chemical control measures, the main safeguard is crop rotation. Some modern

26

varieties of potato are, however, resistant to one of the two species of potato eelworm.

Blight The first sign of attack is the development of yellowish-brown patches on the lower leaves. These quickly turn black and rotten in the moist conditions which favour the spread of the disease. In severe attacks the tubers also become infected and eventually develop an evil-smelling rot. At the first sign of attack the crop should be sprayed with a copper, thiram or zineb fungicide and repeat treatments given at intervals of 7–10 days. Finally, in order to reduce the risk of tuber infection, the haulms should be removed some days before lifting the crop.

Common scab Shows as slightly raised surface scabs on the tubers. These are unsightly but do not affect either the keeping or the eating qualities of the potatoes. The incidence of this disease can be reduced by digging in plenty of bulky organic matter such as farmyard manure, garden compost, peat or composted bark and also by not using lime. In areas where this disease is severe, cropping should be restricted to scab-resistant varieties.

Corky scab The early symptoms of this disease are similar to those of common scab. Later, however, the scabs burst open releasing powdery spores which contaminate the soil. Infected tubers should be burned and potatoes should not be grown on the infected site for at least 3 years.

Gangrene This storage rot is caused by soil-borne organisms which can only infect damaged tubers. Infected tubers become hollow and completely rotted. The incidence of this disease can be reduced by careful lifting and by good storage.

Internal rust spot This appears as scattered brown spots in the flesh of the tubers. Its cause is unknown so no remedial action is possible.

Potato black leg This is a bacterial disease which affects the lower parts of the stems, causing yellowing of the foliage and later the collapse of the haulm. Any tubers which are formed may also be infected and it is these lightly diseased tubers which carry the disease over from one season to the next. The best insurance is to buy fresh, certified seed each year.

Spraing When tubers which are infected with this virus are cut across, narrow reddish-brown curving bands of diseased tissue are clearly visible. The virus is carried by free-living eelworms in the soil, so use fresh seed and carry out crop rotation.

Wart disease This is only a problem on older cultivars as newer varieties are resistant to this disease. Infected tubers are readily recognized by the presence of large warty growths near the eyes. The occurrence of this disease must be notified to the Ministry of Agriculture and, once it is established, only immune varieties can be grown.

PUMPKIN AND SQUASH *See* Marrow.

RADISH
This quick-growing crop is usually trouble-free.

RHUBARB
Crown rot This bacterial disease causes rotting of the terminal bud and of the upper part of the root. It can be troublesome in wet soils. Infected plants carry only small spindly sticks and these, too, may rot. There is no effective control so diseased roots should be burned and new roots established on fresh ground.

Honey fungus This is only a problem on ground which is heavily infected with this root parasite. Where it does occur it can be lethal to the rhubarb. Its presence can be detected by the presence of white fungal threads within the dead root. The only answer to this problem is to make a new planting on a fresh site.

SALSIFY
White blister Shows as white glistening spots on the leaves

making them look unsightly. This disease is favoured by over-crowding, so early thinning reduces the risk of attack. In addition, infected leaves should be removed.

SCORZONERA *See* Salsify.

SEAKALE
Black rot This bacterial disease is most prevalent in warm, wet weather. The leaves of infected plants become yellowed and have dark veins. Cross sections of leaf stalks, stems and roots show dotted rings. Destroy infected plants and practise crop rotation.

Club root *See* Brassicas.

Violet root rot *See* Carrot.

SHALLOT
Mostly trouble-free but may be attacked by onion pests and diseases.

SPINACH
Aphids Blackfly can be troublesome but are readily controlled by greenfly killers or general insecticides.

Downy mildew This shows as a greyish mouldy growth on the lower leaf surface and as yellow patches on the upper surface. Later the infected patches dry out to brownish areas. The disease can be checked by spraying with zineb. It is also helpful to thin out the plants.

Leaf spot This shows on the leaves as brown spots with darker margins. Remove infected leaves and, if necessary, thin out the plants.

Spinach blight This is caused by cucumber mosaic virus. The young leaves become yellow and then later the older ones also change colour. New leaves are narrow and puckered and have inrolled margins. Early control of aphids reduces the spread of the disease.

Boron deficiency *See* Beetroot, heart rot.

SPINACH BEET *See* Beetroot.

SWEDE *See* Turnip.

SWEET CORN

Frit fly The maggots of this fly feed inside the base of the seedling. Emerging leaves are ragged and the shoot may be killed. The seedlings can be protected from attack by the application of HCH dust. Alternatively, the seeds can be germinated indoors and not planted out till late May.

Smut Large galls (smut balls) develop on the ears and stalks. These have smooth, white skins and later open to release a mass of black spores. Remove the smut balls before they open to prevent carry-over of the disease, and practise crop rotation.

TOMATO

Aphids These can sometimes be a problem but are readily controlled by the application of greenfly killers or general insecticides.

Glasshouse whitefly These tiny, white, moth-like insects weaken the plants by feeding on the sap and also deposit large quantities of sticky honeydew which becomes covered with sooty mould. Whitefly are difficult to control because of the resistance of the eggs and larval stages to most insecticides, but they are best dealt with by a programme of three or more sprays of bioresmethrin, pirimiphos-methyl or resmethrin, applied at intervals of 3–4 days.

Blight Potato blight can attack tomatoes causing a browning of the foliage and a black rot of the fruits. Spray with a copper, thiram or zineb fungicide at the first signs of attack and give repeat sprays at intervals of 10–14 days.

Buck-eye rot This disease develops on the fruit as concentric brown rings around a central grey area. It is caused by a

30

soil-borne fungus and infection occurs as a result of splashing from the soil on to the lower fruit. Remove and destroy infected fruit.

Foot rot This shows first as a browning of the base of the stem and can lead to the death of the plant. It can be prevented by raising the plants in sterile growing compost. Badly infected plants should be dug up and burned.

Grey mould A brownish fungal growth develops on rotting stems, fruit and leaves. Remove infected parts and spray the plants with captan, thiram or zineb. In the greenhouse good ventilation should be maintained and care taken with watering to reduce the risk of attack.

Ghost spot This shows as whitish transparent rings on the skin of the fruit. It is caused by grey mould fungus but is not serious as it does not affect the eating quality of the fruit.

Stem rot This is a disease of mature plants and takes the form of a brown canker at the base of the stem. Black dots later develop on this canker. There is no cure, so infected plants should be destroyed and the bases of the remaining plants sprayed with benomyl.

Tomato leaf mould Brownish-purple patches develop on the underside of the leaves followed by yellowing of the upper surface. This only affects greenhouse-grown tomatoes and is favoured by poor ventilation. Control by increasing the ventilation and by spraying the plants with benomyl or thiram.

Virus diseases There are various virus diseases which can infect tomatoes causing curling and mottling of the leaves, thin distorted growth or dark streaking of the stems. There is no cure and infected plants should be destroyed.

Magnesium deficiency This first shows as yellow between the veins on the lower leaves and then spreads upwards. The yellow areas may later turn brown. It can be prevented by the regular use of liquid feeds containing magnesium.

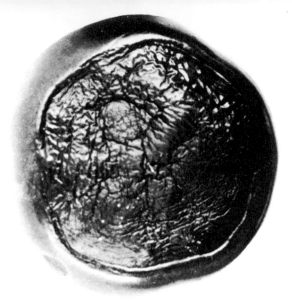

Tomato fruit affected by blossom-end rot

Alternatively, spray with Epsom salts. Dissolve 19g in 1 litre (3oz per gal) and apply every 2 weeks.

Blossom end rot Affected fruits show a circular brown or greenish-black sunken patch at the blossom end. This trouble is caused by irregular watering in the early stages of fruit development.

TURNIP
Cabbage mealy aphid *See* Brassicas.

Cabbage root fly *See* Brassicas.

Flea beetle *See* Brassicas.

Turnip gall weevil The grub of this pest enters the roots causing hollow swellings or galls. It is controlled by applying dusts or granules of bromophos, chlorpyrifos, diazinon or pirimiphos-methyl to the soil at seeding.

Club root Also called 'Finger-and-Toe', this soil-borne disease causes the roots to become swollen and distorted with consequent stunting of growth. The fungus is favoured by acid conditions, so the application of lime before sowing is recommended. The open seed drills should also be treated with calomel (mercurous chloride) dust.

Downy mildew Infected leaves show yellowish blotches and have a brownish mould growth on the underside. This disease mainly affects seedlings and can be checked by spraying with thiram or zineb.

Powdery mildew This appears as a white powdery covering on the leaves in late summer. Because of the lateness of attack no control action is usually needed.

Soft rot This bacterial rot can develop both during growth and in storage. Preventative measures include crop rotation, careful manuring and the avoidance of mechanical and pest damage.

White blister *See* Brassicas.

Turnip mosaic virus The young leaves become twisted and mottled and have dark green raised spots. The aphids which transmit the disease should be controlled by spraying with a greenfly killer or general insecticide.

Brown heart This trouble, which is caused by a deficiency of boron in the soil, shows as greyish-brown rings in the flesh of the root. Affected tubers taste bitter and stringy. The remedy is to apply borax to the soil at the rate of 34g per 20sq m (1oz to 20sq yd).

3 Fruit

APPLE

Aphids Several types of greenfly attack apples. These pests overwinter as eggs on the bark and hatch out when the flower buds begin to grow. All cause damage by feeding, whilst the rosy leaf-curling aphid also produces severe curling and yellowing of the leaves. Winter washes are active against the eggs but spring sprays are essential for good control (Table I).

Woolly aphids, which feed on the bark and are covered with white woolly wax, also overwinter on the trees, becoming active in the spring. They are not controlled by winter washes but can be dealt with by forceful sprays of general insecticide applied after petal fall.

Apple sawfly The larvae of this pest appear around the time that the petals fall. They scar the outside of the fruitlets and then bore their way into the centre of the fruit. For control see Table I.

Apple sucker This pest overwinters as eggs on the bark. The young suckers, which appear in April and May, feed on the leaves and the flowers, often discolouring the latter. They are controlled by winter washes and by spring sprays (Table I).

Capsid These sucking insects feed on the leaves producing pin-point spots. Later, the damaged leaves become tattered, puckered and distorted. Control by DNOC and by spring sprays (Table I).

34

Brown rust on broad bean

Caterpillars of large white butterfly *(Pieris brassicae)* on cabbage

Codling moth The caterpillars of this moth emerge in mid-summer and tunnel into the core of the fruit. For control see Table I.

Fruit tree red spider These tiny pests, which are only just visible to the naked eye, feed on the underside of the leaves causing the leaves to discolour and dry up. Overwintering eggs are killed by DNOC/petroleum but better results are obtained from summer sprays containing derris, dimethoate, fenitrothion, formothion or pirimiphos-methyl. Dinocap, applied to control powdery mildew, also helps to control this pest (Table I).

Tortrix moth The young caterpillars hibernate in cocoons on the tree and emerge during flowering. They bore into the fruit buds and feed on the foliage, spinning leaves together before pupating. A new generation appears later in the summer and these caterpillars feed on both the leaves and the fruit before going into hibernation. For control see Table I.

Winter moth Overwintering eggs hatch at bud burst and the young caterpillars then start to feed on the leaves, flowers and fruitlets. Control by winter washes, coupled with spring sprays (Table I).

Apple scab Infected leaves show olive-green blotches whilst fruits develop brown or black scabs. Control by repeat sprays of benomyl, captan, sulphur or thiram (Table I).

Blossom wilt This fungus attacks the blossom causing wilting. This is followed by die-back of the branch. Where this disease is prevalent, benomyl should be used for the pre-blossom sprays (Table I).

Canker This disease shows as sunken patches on the bark. Branches completely girdled by the canker are killed. Treat by cutting out the brown diseased tissue and then painting the cut surface with a canker paint. Controlling apple scab and woolly aphid, together with the use of canker paint on large pruning cuts, reduces the risk of canker appearing.

Fire blight This bacterial disease causes the rapid browning of the leaves on individual branches and the subsequent die-back of the affected branches. Suspected attacks should be reported to the Ministry of Agriculture who will advise on treatment.

Honey fungus This soil-borne fungus is a common cause of the sudden death of apple trees. It can be identified by stripping pieces of bark from the base of the trunk to expose the whitish, fan-shaped, fungal growths. Brownish 'bootlaces' grow out from infected roots. Dead trees should be dug out together with as much root as possible. The surrounding soil should then be sterilized with 2% formalin using 27 litres per sq m (5gal per sq yd). Alternatively a proprietary armillaria treatment can be used.

Powdery mildew This disease overwinters in the buds and these give rise to shoots, leaves and flowers which are covered with a grey powder of fungal growth. The disease continues to spread to new leaves throughout the summer and also infects the new buds. Pruning gives some control but it is also necessary to remove infected flower trusses and shoots in spring. Complement this treatment by spraying with benomyl, dinocap, sulphur, thiophanate-methyl or thiram (Table I).

Bitter pit This physiological trouble causes surface pitting of the fruit and also small brown spots in the inner flesh of the apple. It is associated with a shortage of water and of calcium. Watering in dry weather is helpful but where the disease reoccurs it is good practice to spray the trees with calcium chloride (80g to 20 litres or $3\frac{1}{4}$oz to 5gal). Apply four sprays at intervals of about a fortnight, starting in mid-June.

APRICOT *See* Peach.

BLACKBERRY
Aphids These sucking insects reduce the vigour of the crop. They are readily controlled by greenfly killers or by general insecticides.

38

Table I Apple spray programme

Spray timing	Suitable chemicals		Pests and diseases controlled
	Insecticides	Fungicides	
December – January	Tar oil		Aphid eggs Apple sucker eggs
January – February	DNOC/petroleum		Capsid eggs Red spider eggs Winter moth eggs
Green cluster – Green flower buds showing			Aphids Apple sucker Tortrix Winter moth Blossom wilt Powdery mildew Scab
Pink bud – Flower buds coloured pink			Aphids Fruit tree red spider Tortrix Powdery mildew Scab
Petal fall – When most of the petals have fallen	dimethoate fenitrothion formothion malathion pirimiphos-methyl trichlorphon	benomyl captan* dinocap† sulphur thiophanate-methyl thiram	Apple sawfly Capsid Fruit tree red spider Tortrix Powdery mildew Scab
Fruitlet (1) – mid-June			Codling moth Fruit tree red spider Tortrix Powdery mildew Scab
Fruitlet (2) – Early July			Codling moth Fruit tree red spider Tortrix Powdery mildew Scab

* Scab only   † Powdery mildew only

Raspberry beetle The grubs of this small beetle feed on the developing fruit. Spray with a general insecticide as the first flowers begin to open.

Crown gall This bacterial disease produces round galls on the lower part of the cane. It is favoured by wet soil. Infected plants should be destroyed and replanting done on better drained soil.

CHERRY
Cherry blackfly Overwintering eggs hatch out at bud burst and the young aphids attack the developing shoots, curling the leaves and stunting the growth. The eggs can be killed by a tar oil spray applied in December. Spraying with a systemic greenfly killer such as dimethoate, fenitrothion, formothion, menazon, pirimicarb or pirimiphos-methyl just before flowering is also effective.

Cherry bacterial canker This bacterial disease enters wounds and leaf scars in the autumn causing the death of the young wood. Later in summer it can cause cankers and the death of whole shoots. Control by applying a copper fungicide at intervals of 3 weeks from the end of August till leaf fall.

CURRANTS, RED, BLACK AND WHITE
Aphids Several types of greenfly attack currants. Some cause a red or yellow blistering of the leaves. Others distort and stunt the leaves and shoots. Overwintering eggs can be killed by the use of tar oil or DNOC/petroleum spray in January. The active aphids are readily controlled by greenfly killers or general insecticides.

Big bud gall mite This major pest of black currants feeds inside the buds which become large and spherical. Infested buds either produce small flowers or fail to open. The mites disperse from these buds in spring and then infest the newly formed buds. Control by removing infested buds in February or March and by spraying with 0.5–1% lime sulphur when the first flowers open. A repeat spray should be given 3 weeks later. Some varieties, such as Amos Black, Davison's Eight,

Goliath and Wellington XXX are 'sulphur shy' and should only be treated with 0.5% lime sulphur.

Black currant leaf midge The tiny grub attacks the foliage causing the leaves to become twisted and folded and the shoot growth to be stunted. Control with carbaryl, dimethoate or pirimiphos-methyl.

Black currant sawfly Attacks by this green, black-spotted caterpillar start in the centre of the bush and progress outwards. Spray with carbaryl, derris, fenitrothion, malathion, pirimiphos-methyl or trichlorphon, taking care to thoroughly spray the centre of the bushes.

Capsid These bugs feed by puncturing the leaves and shoots. Attacked leaves become puckered and torn. Control with carbaryl, dimethoate, fenitrothion, formothion, pirimiphos-methyl or trichlorphon.

Black currant leaf spot With this disease the leaves first develop brown spots. These later run together and the whole leaf turns brown and falls early. Control by spraying with benomyl, thiophanate-methyl, thiram or zineb at fortnightly intervals.

Black currant rust Patches of yellow spores develop on the underside of the leaves in early summer. The leaves later turn brown and fall early. Spray after picking with a copper fungicide or with zineb.

Coral spot Red currants are very liable to be attacked by this disease which causes whole branches to die back. It can be recognized by the presence of coral-red clusters of spores on the dead branches. Prune off the dead shoots and paint the cut surfaces with canker paint.

Powdery mildew Black currant shoots are sometimes affected by powdery mildew late in the season. This disease shows as a white powdery coating on leaves and stems. Control by repeat sprays of benomyl, dinocap or thiophanate-methyl.

Reversion The mature leaves on infected plants are narrower than normal and have only five pairs of veins on the main leaf lobe. Their cropping is reduced and they should be replaced by healthy stock. This virus disease is spread by the big bud mite so it is important to control this pest.

FIG
Coral spot Shoots attacked by this disease die back and later develop coral-red clusters of spores. Infected branches should be sawn off completely and the cut surfaces painted with canker paint.

GOOSEBERRY
Aphids Various species of greenfly attack the bushes. They are readily controlled by greenfly killers or general insecticides.

Gooseberry sawfly The caterpillars of this pest start feeding on the leaves in the centre of the bush and then work outwards. Control with carbaryl, dimethoate, fenitrothion, formothion, pirimiphos-methyl or trichlorphon but care is needed to spray to the centre of the bush.

American gooseberry mildew This disease shows as a powdery white covering on the young leaves, shoots and buds. Later, this covering turns brown. Control the spread of the disease by spraying with benomyl, dinocap or thiophanate-methyl at fortnightly intervals. Any diseased shoot tips should be pruned off and destroyed by burning in late summer.

Gooseberry cluster cup rust This appears in early summer as red or orange swellings on the leaves and fruits. Later, these blisters are covered with minute spore-bearing pits. This disease can be controlled by copper fungicides but these should not be used on dessert varieties.

HAZELNUT, COBNUT AND FILBERT
Nut weevil The female lays her eggs in the young nuts and the grubs then feed on the kernels. There is not, at present, any

42

recommended control treatment available to amateur growers.

Winter moths Caterpillars of this insect feed on the foliage. Control by spraying with a general insecticide at the first sign of attack.

Twig canker This disease attacks cobnuts and filberts. Infected buds turn brown and die in the spring, and then later the fungus enters the twigs, forming cankers. There is not, at present, any effective cure for this common disease.

LOGANBERRY
Aphids These sucking insects weaken the plants. They are readily controlled by greenfly killers or general insecticides.

Raspberry beetle The grubs of this beetle feed on the ripening fruit. Spray with a general insecticide when flowering is nearly over. A further spray with derris can be given when the fruit is starting to colour.

Spur blight This disease first shows in August as dark purplish blotches on the canes. Later, these blotches turn silvery and develop black, pin-point spots. Diseased canes should be cut out. The disease can be prevented by spraying with benomyl, captan, dichlofluanid or thiram, starting when the new canes first develop and then giving repeat treatments at fortnightly intervals.

Virus diseases Symptoms of these diseases are mottling and distortion of the leaves. Destroy infected plants and replant with healthy certified stock.

MEDLAR
Mildew This shows as a white powdery covering on the leaves and as a silvering of the shoots. For control see 'Apple' powdery mildew.

Leaf blotch This disease causes large brown patches on the leaves. Fruits may also be infected and they then become mummified. There is no information on control measures.

MULBERRY

Coral spot Shoots affected by this disease die back and later become covered with coral-red clusters of spores. Infected branches should be pruned off and the cut surfaces painted with canker paint.

NECTARINE *See* Peach.

PEACH

Aphids Some species overwinter as eggs on the trees whilst others invade during the summer. The eggs can be killed with a tar oil spray applied in December. Summer attacks are usually readily controlled by greenfly killers or general insecticides.

Red spider mite These tiny pests, which are only just visible to the naked eye, feed on the underside of the leaves. The first sign of attack is a pale speckling of the leaves which later become discoloured and may be killed. Control by repeat sprays of derris, dimethoate, formothion, malathion or pirimiphos-methyl.

Scale insects The adult females, which stay fixed to the plant, are readily recognized by their shell-like covering. The females die after laying their eggs but the scale remains attached to the branch to act as a cover for the eggs. Young 'crawlers' hatch out from the eggs in summer and move around the tree before settling down to feed. Control by the use of a tar oil wash in early December, or by repeat sprays of malathion or pirimiphos-methyl applied in early summer at 14-day intervals.

Peach leaf curl Infected leaves become swollen and distorted. In the early stages these blisters may be reddish but later they turn white as the spores develop. Control by spraying with 3% lime sulphur or with a copper fungicide in January and again in early February. An alternative is to spray with thiram when the first leaves appear, giving two or three further sprays at fortnightly intervals.

Peach powdery mildew This disease attacks the opening buds, stunting the growth and covering the young leaves with a mealy powder. Later-developing leaves, shoots and fruits are also attacked. Control by repeat sprays of benomyl, dinocap or sulphur.

PEAR
Aphids Several species attack pears but the most important is the pear-bedstraw aphid. These pinkish aphids hatch out from eggs at the white flower bud stage, when they feed on the leaves causing them to curl. The eggs can be controlled by tar oil winter wash, but this treatment needs to be reinforced by pre-blossom sprays with a greenfly killer or general insecticide (Table II).

Codling moth These caterpillars are less common as pests of pears than they are of apples. A single spray with a general insecticide at petal fall is usually sufficient to control these pests.

Capsid These sap-sucking bugs feed by piercing the leaves, producing pin-point spots of damage. Later, the damaged leaves may become tattered and torn. Control by spraying with a general insecticide at petal fall (Table II).

Pear leaf blister mite This microscopic pest overwinters under the bud scales and then invades the leaves in spring. Infested leaves become covered with yellowish or reddish blisters whilst darker coloured pustules develop on the fruits. This is mainly a pest of wall trees. Control by spraying with 5% lime sulphur at the end of March just as the buds start to open. Any infested leaves should also be picked off.

Pear midge This pest is very localized and usually attacks the same tree each year. The small maggots eat into the fruits which blacken and fall early. Control by collecting and destroying the fallen fruit and by cultivating the soil under the tree. General insecticides applied at the white bud stage also have some effect.

Table II   Pear spray programme

Spray timing	Suitable chemicals		Pests and diseases controlled
	Insecticides	Fungicides	
December – January	Tar oil		Aphid eggs Pear sucker eggs
Green cluster – Green flower buds showing	dimethoate fenitrothion formothion malathion pirimiphos- methyl trichlorphon	benomyl captan sulphur thiophanate- methyl thiram	Aphids Winter moth Pear scab
White bud – Flower buds coloured white			Aphids Winter moth Pear midge Pear scab
Petal fall – When most of the petals have fallen			Aphids Capsid Codling moth Pear scab
Fruitlet (1) – mid-June			Codling moth Pear sucker Pear scab
Fruitlet (2) – early July			Codling moth Pear scab

Pear sucker   This sucking pest infests the blossom buds and fouls the foliage with honeydew. There are several generations and the infestation can persist into late summer. Control with general insecticides applied at the fruitlet stage (Table II).

Winter moth   Eggs, which are laid on the bark in winter, hatch at bud burst and the caterpillars then start feeding on leaves, flowers and fruitlets. Winter washes give some control but

46

need to be reinforced by pre-blossom sprays with a general insecticide (Table II).

Canker This disease shows as sunken patches on the bark and, where the branch is completely girdled by the canker, the part above the canker dies. Treat at an early stage by cutting out the brown diseased tissue and then painting the wound with a canker paint.

Fire blight This bacterial disease causes the rapid die-back of individual branches following infection of the flowers. Suspected attacks must be reported to the Ministry of Agriculture who will advise on treatment.

Honey fungus This soil borne fungus can kill pear trees. It can be identified by stripping pieces of bark from the base of the trunk to expose the whitish, fan-shaped, fungal growths. Brownish 'bootlaces' grow out from infected roots. Dead trees should be dug up together with as much root as possible. The surrounding soil should then be sterilized with 2% formalin using 27 litres per sq m (5gal per sq yd). Alternatively a proprietary armillaria treatment can be used.

Pear scab This fungal disease produces olive-green blotches on the leaves and blackish scabs on the fruit. It also attacks young branches which become blistered and scabby. Control by repeat sprays of benomyl, captan, sulphur and thiram (Table II).

Pear stony pit virus This virus causes the fruits to become pitted and deformed. Patches of dead stony cells are also present in the flesh of the fruit so that they are inedible. This disease usually develops on a single branch at first, but later spreads over the tree. There is no control and affected trees should be destroyed.

PLUM

Aphids The leaf curling aphid attacks in spring and the mealy plum aphid in summer. Overwintering eggs can be killed with tar oil in late December or with a DNOC/petroleum wash applied in late January or early February. An alternative approach is to spray with a greenfly killer or general insecticide at white bud and again at petal fall.

Fruit tree red spider These tiny pests, which are only just visible to the naked eye, feed on the underside of the leaves causing them to become discoloured and, later, to dry up. Overwintering eggs can be killed with a DNOC/petroleum wash applied in February. Alternatively, derris, dimethoate, fenitrothion, formothion or pirimiphos-methyl can be applied at petal fall and then repeated as necessary.

Plum sawfly The caterpillars of this pest gnaw black messy holes in the fruit which then fall early. Control with general insecticides applied about a week after the petals have fallen. Regular cultivation of the soil under the trees is also beneficial as it reduces the survival of the pupae.

Tortrix moth Caterpillars of this pest feed on the leaves and also tunnel into the shoots from April to June. Control with general insecticides applied at the white bud stage.

Winter moth The caterpillars of this pest attack the foliage, flowers and fruitlets. They can be controlled by the treatments recommended for use against tortrix.

Bacterial canker This serious disease first shows as brown, rounded spots on the leaves. These brown areas later drop out, leaving holes. Elongated cankers, which exude gum, develop on the branches. Buds on severely affected branches either fail to open or else produce only small yellow leaves. Finally the affected branches die back. The spread of the disease can be checked by cutting off and burning all infected branches. It is also helpful to spray the trees with a copper fungicide in mid-August and to follow this up by repeat treatments in the autumn, say in the middle of September and October.

Blossom wilt and brown rot These diseases cause the blossoms to wilt and are also responsible for a brown rot of the fruit. In addition, they may invade the shoots, causing cankers. Diseased flower trusses and cankered shoots and spurs should be cut off and burned. Mummified fruit should also be collected and destroyed. The dormant season tar oil washes used against aphid eggs also give some control.

Plum rust Bright yellow spots appear on the upper leaf surface and orange or brown spores are produced on the under surface. Later, the infected leaves turn yellow and fall early. Carry-over of the disease to next year can be reduced by sweeping up and burning the infected leaves. Keep a close watch for recurrence of the disease and spray with thiram at the first signs of attack.

Silver leaf This fungal disease shows up first as a silvering of the leaves on infected branches. Later there is a progressive die-back of the affected branches and small, bracket-shaped, fruiting bodies may develop on the dead wood. Cross sections of infected branches show a purplish staining of the wood. Control by cutting back affected branches to 15cm (6in) below the level of the stained wood and then paint the cut surface with a canker paint.

False silver leaf This physiological trouble can easily be mistaken for true silver leaf. The foliage is silvered but most of the leaves on the tree are affected at the same time, and there is no die-back of the branches, nor is there any staining of the wood. This trouble is caused by malnutrition and by variations in the water supply, and it can be corrected by feeding, mulching and watering as necessary.

QUINCE
Leaf blight and fruit spot This disease shows on young leaves and shoots as irregular red spots. These later coalesce and infected leaves may fall early. Diseased fruits have dark brown sunken spots and may also be deformed. Control by pruning off infected branches and then applying a copper fungicide.

Powdery mildew *See* Apple.

RASPBERRY

Aphids These pests not only weaken the plants but also transmit virus diseases. They are readily controlled by greenfly killers or general insecticides.

Raspberry beetle The small grubs, which hatch out from eggs laid on the blossoms, feed on the developing fruit. Control by spraying with derris, fenitrothion or malathion when the first pink fruit are seen.

Cane blight This fungal disease attacks the base of the canes. Dark areas then develop on the base of the cane which becomes very brittle. Diseased canes should be cut back to below soil level and the new canes protected by spraying with a copper fungicide.

Cane spot This first shows in May or June in the form of purple spots on the canes. These spots later elongate and become whitish with a purple border. Later, the centres of the spots split, giving the cane a ragged look and in severe cases the tips of the canes may die. Spots may develop on the leaves and infected fruits become distorted. Control by spraying with benomyl, dichlofluanid or thiram at fortnightly intervals from bud burst till pink bud.

Spur blight This fungus infects the new canes in June but it is not until August that the typical dark purple blotches are seen on the canes at the leaf joints. These blotches later enlarge and turn silvery. Buds at the infected nodes either die or produce shoots which soon die back. Infected canes should be removed as soon as possible. Control by spraying with captan, dichlofluanid, thiophanate-methyl or thiram. The first spray should be given when the new canes are quite short and repeat sprays should then be applied at fortnightly intervals over the next 2 months.

Virus diseases The common symptoms of these diseases are yellow mottling, blotching and distortion of the leaves. In addition, the plants may be stunted and the cropping reduced. There is no cure and infected stools should be lifted and

50

burned, new plantings being made with certified healthy canes on a fresh site.

STRAWBERRY

Aphids These are common pests on strawberry plants. They are relatively easy to control with greenfly killers or general insecticides.

Strawberry beetle These pests eat the seeds and also damage the fruit. They can be controlled by the use of slug pellets based on methiocarb.

Strawberry mite These colourless mites, which are only just visible to the naked eye, feed on the young folded leaves. Infested leaves remain small and later turn brown whilst heavy attacks cause the plants to be stunted. Control by spraying with dicofol after picking the crop.

Slugs These can be very damaging to the fruit and should be controlled with slug pellets.

Grey mould This disease causes a brownish-grey rot of the fruit following an earlier infection of the flowers. Control by spraying with benomyl, captan, dichlofluanid or thiram at early blossom and, if necessary, applying repeat sprays at fortnightly intervals.

Leaf spots and blotch Several fungi can cause the leaves to become spotted but these attacks are not serious. Leaf blotch, which shows as brown blotches with purple margins, is more damaging because it can cause rotting of the leaf and flower stalks. Generally, it is sufficient to simply remove any infected leaves. If, however, the trouble persists, the plants should be sprayed with zineb or dichlofluanid when growth starts in the spring and again just before flowering.

Strawberry mildew This disease shows up as purplish patches on the upper surface of the leaves whilst the undersides have a greyish covering of fungal spores. Fruits may also be attacked causing them to shrivel. Control by spraying with benomyl, dinocap, sulphur or thiophanate-methyl at 10–14-day intervals.

Virus diseases These generally cause stunting and poor cropping but they can also kill the plants. Infected plants should be dug up and burned. New plantings of fresh, healthy runners should be made on a new site.

VINE, OUTDOOR
Aphids These common pests are readily controlled by greenfly killers or by general insecticides.

Downy mildew This shows as light green patches on the upper surface of the leaves, with corresponding areas of fungal growth on the underside. These diseased areas later become dry and brittle. The berries, too, can be affected and in severe cases this can cause them to shrivel. Control by repeat sprays with a copper or zineb fungicide.

Powdery mildew This disease is common both under glass and outdoors. It shows as a white powdery deposit on both the leaves and the fruit. Infected berries later become discoloured and may split. Control by repeat treatments with dinocap or sulphur.

4 Trees and Shrubs

These can be affected by a wide range of different pests and diseases but are all liable to be attacked by the soil-borne honey fungus (*Armillaria mellea*). This usually shows initially as an out of season yellowing and collapse of the foliage. The first sign of attack, however, may be the failure of the tree or shrub to come into leaf in the spring. Later, the whole plant dies back. *Armillaria* can be distinguished from other diseases causing premature leaf fall and die back by the presence of white, fan-like, fungal growths under the bark at the base of the trunk. The existence of this growth can easily be checked by removing the top layer of soil around the trunk and then chipping off small pieces of bark at intervals around the collar.

Dead trees or shrubs should be dug up and burned together with as much root as possible. Any brown-black fungal strands (rhizomorphs) growing from these roots, or in the soil generally, should also be removed. When replacing the soil each 10–15cm (4–6in) layer should be soaked with a solution of tar oil emulsion designed for use against *Armillaria*. This treatment is needed to kill any remaining fungal strands in the soil and also to prevent the development of the tawny-coloured toadstools which are responsible for spreading the disease. Nearby trees and shrubs can be further protected from attack by the application of heavy soil drenches of the special tar oil emulsion.

ACER (MAPLE)
Horse chestnut scale These sucking insects can be recognized by their brown, shell-like covering. The females produce a

mass of eggs, covered with white waxy wool, in May or June, and then die shortly before the eggs hatch. The young larvae then crawl to suitable feeding sites before settling down and producing their protective scales. Control by spraying with malathion or pirimiphos-methyl in June.

Pocket galls These show up as rounded pimples on the upper surface of the leaves. They cause little damage and no control action is needed.

Tortrix The small caterpillars feed inside clusters of leaves which they roll up and tie together with webbing. Some control can be obtained by repeat sprays of trichlorphon or pirimiphos-methyl.

Tar spot fungus The black, tarry spots of this fungus commonly appear on the leaves of maples and sycamores in late summer but the disease has no serious effects on the trees.

ALMOND
Caterpillars Caterpillars of various moths and sawflies can seriously damage the foliage. Control with general insecticides applied at the first sign of damage.

Fruit tree red spider These tiny pests, which are only just visible to the naked eye, feed on the underside of the leaves causing discoloration of the foliage. Control by repeat sprays of dimethoate, formothion, malathion or pirimiphos-methyl.

Peach leaf curl Leaves infected with this fungus become swollen, blistered and deformed. They fall early. Control by applying a copper fungicide or 3% lime sulphur in January and again in early February. A third spray should be applied in autumn before leaf fall. Alternatively, three sprays of thiram can be applied at 10 day intervals in early summer.

AZALEA
Azalea leaf miner The grubs of this pest tunnel into the leaves producing blister-like mines. Control by repeat sprays of dimethoate, malathion, pirimiphos-methyl or trichlorphon.

54

Azalea whitefly These pests, which look like tiny white moths, feed on the leaves, making them look unsightly with honeydew and sooty moulds. Control by a programme of three or four sprays of bioresmethrin, pirimiphos-methyl or resmethrin applied at intervals of 3–4 days.

Caterpillars Leaf-eating caterpillars can be troublesome. These pests are readily controlled by general insecticides.

Vine weevil The small wingless beetles feed on the leaves producing a characteristic marginal notching. Control by spraying or dusting the foliage and the ground underneath the bush with a general insecticide.

Azalea gall This fungus disease of evergreen azaleas shows up as round galls on the leaves and flower buds. The galls are reddish when young but turn white when the spores develop. Pick off any galls and then give repeat sprays with a copper fungicide, captan or zineb.

BEECH

Beech aphid This common pest, which is covered with white wool, congregates on the underside of the leaves and produces copious quantities of sticky honeydew. Heavy infestations cause the leaves to turn brown and shrivel. This pest may be controlled by the use of greenfly killers or general insecticides.

Beech scale This insect shows up clearly on the bark because of its white, woolly wax covering. Young 'crawlers' hatch out in late summer or early autumn and these can be dealt with by the use of repeat sprays of malathion or pirimiphos-methyl over this period.

Caterpillars Moth caterpillars can sometimes be damaging to the leaves. Control with general insecticides.

Felt galls These appear as swellings on the upper surface of the leaves with corresponding hollows on the underside. No control action is necessary.

BIRCH
Caterpillars Moth caterpillars, which feed on the leaves, can be controlled by general insecticides. More serious, but less common, are the caterpillars of the goat moth which tunnel into the wood and can kill single branches or even whole trees. This pest is difficult to control because it is not easy to spot before the damage is done.

Leaf weevils These beetles cut holes in the leaves. Control with the use of general insecticides.

Felt galls *See* Beech.

BOX
Box sucker The young larvae of this pest invade the opening buds and cause the leaves to curl, forming a cabbage-like gall at the end of the shoot. They exude white, curled, waxy tubes filled with honeydew which further disfigures the bush. Control by spraying with dimethoate, formothion or pirimiphos-methyl at the first sign of damage. Further sprays should be applied at the end of July and again in mid-August to kill off the adults before they can lay their eggs.

Mussel scale This pest can be recognized by its mussel-shaped scaley covering. Control by repeat sprays of malathion or pirimiphos-methyl applied in early summer when the eggs are hatching.

BROOM
Bud galls These are the result of the buds becoming infested with gall mites. They are composed of swollen bud scales and look like miniature green roses. They are difficult to control and are best picked off and burned.

Willow scale The females appear as small, white, irregular scales whilst the smaller males are rod-shaped. Control by repeat sprays of malathion or pirimiphos-methyl applied in early summer when the eggs are hatching.

BUDDLEIA
Caterpillars Leaf-eating caterpillars can be a problem but are readily controlled by general insecticides.

Clay-coloured weevil These wingless beetles gnaw the bark of young shoots. They also bite holes or notches in the leaves and can gnaw through the stems of leaves and flowers. Control by spraying both the foliage and the soil under the bush with a general insecticide.

Red spider mite These tiny mites, which are only just visible to the naked eye, feed on the underside of the leaves causing them to become yellow and sickly looking. Control by repeat sprays with dimethoate, formothion, malathion or pirimiphos-methyl.

CAMELLIA
Vine weevil These small wingless beetles eat characteristic notches from the leaf margins. They can be controlled by spraying the foliage and the soil underneath the bush with a general insecticide.

CARYOPTERIS
Brown scale The adult scales can be recognized by their rounded, chestnut-coloured, shell-like covering. The eggs are laid in May and hatch out at the end of June, so they are best controlled by spraying in July and August with malathion or pirimiphos-methyl.

Capsid bug Eggs laid on the young shoots in autumn hatch out in April and the young bugs then feed on the leaves, producing calloused spots. Later, the leaves may become torn and ragged. Control by spraying both the plant and the surrounding soil with a general insecticide at the first sign of damage.

CEANOTHUS
Caterpillars These hatch out in early summer from cocoons

on the bark and then feed on the foliage and buds. Control with general insecticides.

Mealybugs These sap-sucking insects, which can be recognized by their protective covering of white powdery wax, are a common pest in south-west England. Control with malathion or pirimiphos-methyl, but a forceful spray is needed to penetrate the waxy covering.

Vine weevil *See* Camellia.

Willow scale *See* Broom.

CHERRY, ORNAMENTAL
Brown scale These pests can be recognized by their rounded, chestnut-coloured, scaley covering. The eggs are laid in May and hatch out at the end of June. They are best controlled by repeat sprays of malathion or pirimiphos-methyl applied in July and August.

Caterpillars Caterpillars of various moths and sawflies feed on the foliage and buds. Control with general insecticides.

Cherry blackfly This pest, which causes severe leaf curl and also stunts young growth, is a common pest of ornamental cherries. It is best controlled by spraying with a systemic greenfly killer, such as menazon or pirimicarb, shortly after bud burst. If necessary, a second treatment can be given after petal fall.

Clay-coloured weevil This wingless beetle feeds at night on the foliage, biting holes in the leaves. Control by spraying the tree and the soil underneath with a general insecticide.

Cherry bacterial canker This bacterial disease gains entry through wounds and leaf scars in the autumn, killing off the young wood. It can also cause cankers which may lead to the death of whole branches. Diseased branches should be cut out and a copper fungicide applied at 3-week intervals from the end of August until leaf fall.

CLEMATIS
Clay-coloured weevil *See* Cherry, ornamental.

Soft scale This pest, which feeds on the leaves, can be recognized by its flat, yellowish-brown, translucent scaley covering. Control by repeat sprays of malathion or pirimiphos-methyl.

Clematis wilt This rather mysterious disease causes the sudden die-back of one or more of the shoots. Fortunately, new shoots generally emerge without remedial action.

Powdery mildew This disease produces a powdery covering of the leaves. It can be controlled by spraying with bupirimate, dinocap, thiophanate-methyl or triforine.

CONIFERS
Adelgids These tiny sucking insects, which only attack conifers, are easily seen because of the white, woolly wax which they produce. They also excrete masses of sticky honeydew on which sooty moulds become established. Some species feed on the young shoots, which swell up into large scaley galls, whilst others feed on the needles. Finally, there are some species which restrict their attack to mature branches, on which they produce swollen tumours. They can attack spruce, douglas fir, Scots pine, Weymouth pine and silver fir. Control of these pests can be obtained by spraying with a general insecticide in late spring or early summer.

Conifer spinning mite These tiny mites can cause serious damage to cedar, cypress, juniper, thuja, pine and spruce. The foliage on infested trees becomes grey or yellowish and may wither and die. Infested shoots may also become covered with a fine webbing. These pests overwinter as eggs which hatch in early May. Control by spraying with dicofol, malathion or pirimiphos-methyl in late May, followed by repeat treatments at intervals of 2–3 weeks.

Yew scale This pest can be identified by its rounded, chestnut-coloured, shell-like covering. Control by repeat sprays of malathion or pirimiphos-methyl in July and August.

COTONEASTER

Brown scale This pest can be identified by its rounded, chestnut-coloured, shell-like covering. In fact it looks very much like yew scale. Control by spraying in July and August with malathion or pirimiphos-methyl.

Caterpillars Various species feed on the leaves of cotoneasters. Control by the use of general insecticides.

Woolly aphid These sucking pests, which feed on the bark, are easy to recognize because of their white, woolly wax covering. They can be controlled by forceful sprays of general insecticides.

Fire blight This bacterial disease causes a rapid die-back of individual branches in the summer, following infection through the flowers. Suspected attacks should be reported without delay to the Ministry of Agriculture who will advise on treatment.

CRAB APPLE   *See* Apple in chapter on 'Fruit'.

CURRANT, FLOWERING

Brown scale *See* Cotoneaster.

Caterpillars Various types of caterpillar can feed on the foliage. Control by spraying with a general insecticide.

Woolly aphid *See* Cotoneaster.

EUONYMUS

Blackfly Overwintering eggs hatch out in spring and the blackfly then infest the young growth, causing leaf curl. They are best controlled by spraying with a systemic insecticide at the first sign of attack.

Willow scale The females show up as white, irregular-shaped scales on the bark whilst the males are rod-shaped. Control by repeat sprays of malathion or pirimiphos-methyl applied in early summer as the eggs are hatching.

FORSYTHIA

Common green capsid Overwintering eggs hatch in April and the larvae then start to feed on the foliage. Their feeding punctures first show up as calloused spots but later the leaves may become ragged and torn. Control by spraying the bush and the ground underneath with a general insecticide.

Mealybugs These sap-feeding pests, recognized by their white powdery wax covering, can be troublesome in south-west England. Control with malathion or pirimiphos-methyl; a forceful spray is needed to penetrate their waxy covering.

FUCHSIA

Aphids Greenfly can be a problem but they are easily controlled by greenfly killers or general insecticides.

Common green capsid *See* Forsythia.

Froghoppers These pests are easily spotted by the frothy 'cuckoo spit' which covers the larvae as they feed in the axils of the leaves. Control by forceful sprays of general insecticides.

Leaf-cutter bees These cut characteristic circular and oval pieces from the leaf margins. There is no easy way to control these pests but luckily they do little harm.

HAWTHORN

Caterpillars The foliage may be damaged by several different types of caterpillar. These pests can be controlled with general insecticides.

Red spider mite These tiny pests, which are only just visible to the naked eye, feed on the foliage causing it to become discoloured. Control by repeat sprays of malathion or pirimiphos-methyl.

Woolly aphid These sucking insects, which feed on the bark, can be recognized by their white, woolly, waxy covering. Control by drenching sprays of general insecticides.

Fire blight This bacterial disease causes the rapid die-back of branches during the summer, following infection of the flowers. Suspected attacks must be reported to the Ministry of Agriculture who will advise on treatment.

Powdery mildew With this disease the leaves and shoots become covered with a white, powdery, fungal growth. Control by repeat sprays of bupirimate, dinocap or triforine.

HEATHER
Mussel scale The adult females look like miniature brown mussels on the bark. Control by repeat sprays of malathion or pirimiphos-methyl applied in late summer and early autumn.

HIBISCUS
Soft scale This pest feeds on the underside of the leaves and can be recognized by its flattened, orange or brown translucent covering. This is normally a pest of greenhouse-grown plants but can sometimes be found out of doors. Control by repeat sprays of malathion or pirimiphos-methyl.

HOLLY
Caterpillars The tender young leaves can be attacked by caterpillars which bind them together with silky webs. Isolated attacks can be dealt with by pruning off the infested shoot tips. Alternatively, the shrubs can be protected by applying repeat sprays of general insecticides.

Holly leaf miner The small grubs burrow into the leaf tissue from early June onwards producing yellowish-brown blisters. Control by repeat sprays of systemic general insecticides.

Horse-chestnut scale This is most easily spotted in May or June when the female produces a mass of white, woolly, wax-containing eggs. Control by repeat treatments with malathion or pirimiphos-methyl, starting when the eggs are first seen and applied at 14-day intervals.

HONEYSUCKLE
Aphids Honeysuckle is commonly attacked by greenfly and

blackfly. Control with greenfly killers or general insecticides.

Brown scale This pest can be identified by its rounded, chestnut-coloured, scaley covering. Control by spraying in July and August with malathion or pirimiphos-methyl.

Honeysuckle whitefly These tiny, white, moth-like insects can sometimes be troublesome. Control with a programme of three or four sprays with bioresmethrin, pirimiphos-methyl or resmethrin, applied at intervals of 3–4 days.

Leaf miner The tiny grubs of this pest burrow into the leaf tissue, producing linear 'mines'. The damage generally does not warrant taking control measures other than picking off the infected leaves.

HYDRANGEA
Capsid bugs These pests feed by puncturing the plant tissue, causing calloused spots. Attacked leaves may later become torn and ragged. Control by spraying the foliage and the ground under the shrub with a general insecticide.

Clay-coloured weevil These wingless beetles feed at night, gnawing the bark of young shoots. They also bite holes or notches in the leaves and can gnaw through the stems of leaves and flowers. Control by spraying the bush and the soil underneath with a general insecticide.

Red spider mite These tiny pests, which are only just visible to the naked eye, feed on the underside of the leaves causing discoloration. Control by repeat sprays of derris, dimethoate, formothion, malathion or pirimiphos-methyl.

Vine weevil This night-feeding beetle bites characteristic deep notches in the margins of the leaves. For control see under 'Clay-coloured weevil'.

Powdery mildew This disease produces brown patches on the leaves with the mildew showing as a whitish covering. Control by repeat sprays of benomyl or dinocap.

IVY
Caterpillars These pests can cause leaf damage. Control with the use of general insecticides.

Horse-chestnut scale *See* Holly.

JASMINE, WINTER
Willow scale The females show up as white, irregular-shaped scales on the bark whilst the males are rod-shaped. Control by repeat sprays of malathion or pirimiphos-methyl applied in early summer when the eggs are hatching.

LABURNUM
Mealybugs These are a common pest in south-west England. They are small sucking insects which are protected by a covering of white powdery wax. Control with malathion or pirimiphos-methyl but a forceful spray is required to penetrate the waxy covering.

Laburnum leaf miner These small grubs burrow into the leaf tissue eventually forming a large, irregular blotch near the edge of the leaf. Damaged leaves become shrivelled and brown. This pest not only disfigures the foliage but can stunt the growth of young trees. Control by repeat sprays with a general insecticide, beginning at the first sign of attack.

LAVENDER
Froghopper These pests can be detected by the frothy 'cuckoo spit' which surrounds the feeding larvae. Control by forceful sprays with a general insecticide.

Shab The first sign of attack by this soil-borne fungal disease is the wilting and subsequent death of some of the shoots. Later the disease spreads to other shoots and causes the death of the plant. There is no cure and the only answer is to make fresh plantings on another site.

LILAC
Caterpillars Various types of caterpillar can attack and damage the foliage. Control with the use of a general insecticide.

Lilac leaf miner The small grub of this pest burrows into the leaf tissue forming a large brown blister. Control by repeat sprays with a general insecticide.

Privet thrips These small sucking insects cause severe silvering and distortion of the leaves. Control with general insecticides.

Willow scale *See* Jasmine, winter.

Blight This bacterial disease shows up as brown spots on the leaves. It also causes serious damage to young shoots which blacken and wither. It is best controlled by pruning off infected shoots and then applying a copper fungicide.

Verticillium wilt Invasion of the tree by this soil-borne fungus causes the leaves to wilt and turn pale before falling early. This is a deadly disease and infected trees should be dug up with as much root as possible and then burned. No further plantings of lilac should be made on the same site.·

MAGNOLIA
Brown scale This bark-feeding pest can be recognized by its rounded, chestnut-coloured scaley covering. Control by spraying in July and August with malathion or pirimiphos-methyl.

MAHONIA
Powdery mildew This disease shows as a white powdery covering on the leaves, especially on the young growth. Control by spraying with dinocap.

Rust Small red spots appear on the upper leaf surface whilst brown powdery spores are produced on the lower surface. Spraying with thiram gives some control of this disease.

MAPLE   *See* Acer.

PRIVET
Lilac leaf miner *See* Lilac.

Privet thrips These sucking insects cause silvering and distortion of the leaves. Control by spraying with a general insecticide.

Willow scale The females show up on the bark as irregular white scales whilst the males are rod-shaped. Control by repeated applications of malathion or pirimiphos-methyl in early summer when the eggs are hatching.

PYRACANTHA
Caterpillars These pests can damage the foliage but are readily controlled by the use of a general insecticide.

Nut scale This bark-feeding pest can be recognized by its rounded, brown scaley covering. Control by spraying with malathion or pirimiphos-methyl in July and August.

Woolly aphid This sucking insect shows up on the bark because of its white, woolly wax covering. Control by forceful sprays of general insecticide.

Fire blight This bacterial disease causes the sudden wilting of the foliage during the summer, following infection through the flowers. Suspected cases must be reported to the Ministry of Agriculture who will advise on treatment.

Scab This disease shows as an olive-brown or black coating of the fruits and leaves. Control by spraying with benomyl, captan or thiram in late spring and early summer.

RHODODENDRON
Caterpillars Various types of caterpillar can attack the foliage and cause damage. Control by spraying with a general insecticide at the first sign of attack.

Rhododendron bug These tiny yellow or brown insects feed in groups on the lower surface of the leaves, producing a characteristic rusty spotting whilst the upper leaf surface becomes mottled with yellow. Control by spraying with a general insecticide at the first sign of attack in June.

66

Rhododendron leafhopper This vividly coloured insect feeds on the foliage and flower buds. It causes little direct damage but is responsible for the infection of the flower buds with the fungus which causes bud blast. Control by spraying with a general insecticide at 3-week intervals in August and September.

Rhododendron whitefly Both the adults, which look like miniature white moths, and the larvae feed on the underside of the leaves, producing honeydew and the associated sooty moulds. Control with a programme of three or four sprays with bioresmethrin, pirimiphos-methyl or resmethrin applied at 3–4-day intervals.

Weevils These small beetles feed at night on the leaves, notching the margins or biting holes in the leaf blade. Control by spraying the plants and the soil underneath them with a general insecticide.

Bud blast This fungal disease makes the flower buds turn brown, black or silvery in the spring. Infected buds fail to open. It enters the buds through wounds caused by rhododendron leafhopper so controlling this pest is the best way of dealing with bud blast disease.

ROSE
Aphids Roses are subject to attack by various species of aphid throughout the summer. These pests are readily controlled by spraying with greenfly killers or with general insecticides.

Capsid bugs These sucking insects feed on leaves and flower buds producing small calloused spots. Severely damaged leaves may later become torn and ragged. Control with the use of general insecticides.

Caterpillars These pests can cause serious damage to the foliage by eating holes in the leaves or by skeletonizing them. Control by the use of a general insecticide at the first sign of attack.

Chafer beetles The adult beetles feed on the foliage and also ruin the blooms by boring into the flower buds. The whitish fleshy grubs, on the other hand, feed on the roots. The adults can be controlled by the use of general insecticides whilst one way of dealing with the larvae is to work HCH dust into the soil around the bush. Alternatively, a heavy soil drench of spray-strength general insecticide can be used.

Clay-coloured weevil These wingless beetles feed at night, gnawing the bark of young shoots. They also bite holes and notches in the leaves and can gnaw through the stems of leaves and flowers. Control by spraying the bush and the ground underneath with a general insecticide.

Leaf-cutter bees These cut characteristic circular and oval pieces from the leaf margins. There is no easy way to control these pests but luckily they do little harm.

Leaf rolling sawfly Infestation by this pest can be detected by the presence of leaflets which are rolled into tight tubes. It is difficult to control because the grubs are protected within the rolled-up leaflets. Infested leaves should be removed and burned and the bushes sprayed with a general insecticide.

Red spider mite These tiny pests, which are only just visible to the naked eye, feed on the lower leaf surface and cause a bronzing of the foliage. Control by repeat sprays of malathion or pirimiphos-methyl.

Rose leafhopper These small, yellow, moth-like insects feed on the lower surface of the leaves causing a pale mottling of the upper surface. Control with general insecticides.

Rose thrips These tiny elongated insects feed on the petals of the opening buds causing them to become distorted and streaked with brown lines. They also feed on the foliage, producing a silvering of it. Control with general insecticides.

Scale insects Various species of these pests, with their distinc-

Damage caused by the leaf-cutter bee to rose foliage

tive shell-like protective scales, can attack roses. Control by repeat sprays of malathion or pirimiphos-methyl.

Black spot This very common and damaging disease shows up as large, circular, black or purple spots on the leaves. It can also develop on the young stems. Control with a programme of fortnightly sprays starting in early spring and continuing throughout the summer, using benomyl, captan, dichlo-fluanid, dicofol, folpet, thiophanate-methyl or triforine. To avoid the risk of the build-up of disease resistance it is recommended that the fungicide used should be changed from time to time.

69

Black spot on rose foliage

Grey mould This fungus can attack opening flower buds, causing the outer petals to turn brown. Affected blooms may stop developing and become 'capped'. This trouble is most common in cold wet periods and there is no effective method of control. Shoots bearing these damaged buds should be pruned back to encourage regrowth.

Leaf scorch This disease first appears as yellowish-green patches on the leaves. Later these patches turn brown and become bordered with a red or purple line. Normally this is not a serious problem but in some years it can reach epidemic proportions, causing complete defoliation before the end of

July. The spray programme used to control black spot can be expected to be effective against this disease also but if control breaks down it is suggested that a copper fungicide should be used.

Powdery mildew Leaves, shoots and flower buds become covered with a white, powdery, fungal growth. Control by spraying with benomyl, bupirimate, carbendazim, dinocap, thiophanate-methyl or triforine.

Rust This is not a common disease in most areas which is fortunate since it can be very damaging. The first sign of attack shows in early spring when small, yellow, spore-bearing spots develop on the underside of the leaves and on the petioles. These are not easy to see and usually the disease is not recognized until mid-summer, when the underside of infected leaves carries numerous orange pustules and the upper surface develops yellow spotting. Still later, the spore colour changes to black and about this time there is rapid defoliation of the infected bushes. Control by repeat applications of maneb, thiram or triforine.

SORBUS
Fire blight This bacterial disease causes the rapid die-back of whole shoots during the summer, following infection of the flowers. Suspected attacks must be reported to the Ministry of Agriculture who will advise on treatment.

VIBURNUM
Blackfly The Guelder rose (*V. opulus*) is a winter host for blackfly eggs. In spring these hatch into blackfly which attack the developing leaves. They are readily controlled by greenfly killers or general insecticides.

Caterpillars These pests can damage the foliage. Control with general insecticides.

Viburnum whitefly These tiny, white, moth-like insects and their distinctive larvae, which are black and have a white waxy fringe, can be very damaging to *Viburnum tinus*. Indeed,

71

severe infestations can cause serious defoliation and even die-back. Control with a programme of three or four sprays of bioresmethrin, pirimiphos-methyl or resmethrin applied at intervals of 3–4 days.

WILLOW

Caterpillars The small grey caterpillars of the willow ermine moth feed in colonies on the foliage enclosed within a web 'tent'. They are best controlled by spraying with a general insecticide in spring before the protective tents have been formed.

Willow scale The females of this pest show up as white irregular scales on the bark whilst the males are rod-like. Control by repeat sprays of malathion or pirimiphos-methyl in early summer.

Willow stem aphid These large, dark brown aphids are exceptional in that they feed on the bark rather than on the foliage. They can be controlled by thorough spraying with a greenfly killer or general insecticide.

Anthracnose This fungal disease shows up as small brown spots on the leaves and as blackish cankers on the stems. Control by spraying with a copper fungicide as the leaves unfold and then giving a repeat treatment in the summer. Heavily infected branches should be pruned off and it is also helpful to apply a general fertilizer over the root zone of the tree.

WISTERIA

Clay-coloured weevil These beetles eat holes in the foliage and may also notch the margins of the leaves. In addition they may bite through the stems of leaves and flowers. Control by spraying both the foliage and the ground underneath the infected plant with a general insecticide.

Brown scale This pest can be recognized by its rounded, chestnut-coloured, scaley covering. Control by spraying in July and August with malathion or pirimiphos-methyl.

5 Outdoor Flowers

ACONITUM (MONKSHOOD)
Delphinium moth The caterpillars of this moth damage the buds, leaves, flowers and seed capsules. Control with general insecticides.

ALTHAEA (HOLLYHOCK)
Caterpillars Foliage-feeding caterpillars are readily controlled by the use of general insecticides. Stem-boring caterpillars, however, present difficulties. These tunnel into the stem, causing the shoot to collapse, and can not be dealt with by sprays. Remove the damaged stems and destroy the caterpillars or pupae which will be found inside.

Rust This extremely common and serious disease shows up as raised pustules on the lower surface of the leaves and on the stems. These pustules are whitish-yellow at first but then turn brown as the spores develop. Heavy infections can kill the plants. Protect seedling plants by spraying with thiram at fortnightly intervals.

ALYSSUM
Flea beetle These beetles feed on the foliage, leaving it pitted with small holes. Control by the application of insecticidal dust to the plants and to the surrounding soil.

ANEMONE
Aphids These pests not only damage the plants by their feeding but also transmit virus disease. Control with greenfly killers or general insecticides.

73

Caterpillars These damage the leaves, stems and buds. Control with general insecticides.

Flea beetle *See* Alyssum.

Downy mildew This disease shows as a whitish mould on the underside of the leaves which tend to roll upwards. Control by repeat sprays of thiram or zineb.

Grey mould This fungus can produce a rot of flowers and flower buds in wet weather. Control by benomyl, captan or thiram.

ANNUALS, GENERAL
Capsid bugs These bugs feed by puncturing the plant tissue causing brown calloused spots on leaves and flowers. Control with general insecticides.

Cutworms These soil-living caterpillars gnaw both the plant roots and the stems, often severing the latter at soil level. Control by raking in a soil insecticide such as bromophos, chlorpyrifos, diazinon, HCH or pirimiphos-methyl. Alternatively, apply a heavy soil drench of spray-strength general insecticide.

Slugs These pests can be very damaging to annuals and to bedding plants in general. Control by the use of slug pellets.

ANTIRRHINUM
Aphids These pests weaken the plants by their feeding and also disfigure the foliage with honeydew and its associated sooty moulds. Control with greenfly killers and general insecticides.

Caterpillars Foliage-feeding caterpillars can be controlled by general insecticides. The stem-boring caterpillars of the rosy rustic moth cannot be controlled by spraying and the only answer is to remove the damaged stems.

Downy mildew This shows as a grey mealy growth on the

74

Rust on antirrhinum leaves

underside of the leaves. Control by repeat sprays of thiram or zineb.

Rust This shows as brown sporing spots on the leaves and can be very damaging. Control by repeat sprays of maneb, thiram or zineb. An alternative approach is to grow rust-resistant varieties.

ASTER, CHINA *See* Callistephus chinensis.

AUBRIETA
Cabbage root fly The whitish maggots of this pest feed on the plant roots, stunting the growth and causing a tendency for the plants to wilt in hot weather. Control by dusting the soil round the plants with bromophos, chlorpyrifos, diazinon, HCH or pirimiphos-methyl.

AURICULA
Root aphid Plants infested with these root-feeding aphids make slow growth and tend to wilt in dry weather. The aphids show up clearly on the roots because of their white waxy covering. Control with heavy soil drenches of spray-strength general insecticide.

BEGONIA
Thrips These minute slender insects feed on the leaves, young shoots and flowers causing a pale-yellow or silvery mottling of the leaves and a white or brown flecking of the petals. They may also deposit honeydew. Control with general insecticides.

Powdery mildew This disease causes white powdery spots on the leaves and stems of bedding begonias. It is difficult to control but repeat sprays of dinocap or thiram may be effective.

BLUEBELL *See* Endymion.

CALLISTEPHUS CHINENSIS
Aphids These pests weaken the plants by their feeding and also disfigure the foliage with honeydew and its associated sooty moulds. Control with greenfly killers or general insecticides.

Capsid bug This pest feeds on the leaves and flower buds producing brown calloused spots. The infested leaves may later tear and become ragged. Control with general insecticides.

Caterpillars These feed on the flowers, tying the petals together to form a shelter. Control by hand picking or with the use of general insecticides.

Cutworms These soil-living caterpillars feed on the roots and on the base of the stem. Control by raking in soil insecticides or by applying soil drenches of spray-strength general insecticide.

76

Chrysanthemum eelworm This microscopic pest invades the leaves causing brown or black patches between the veins. There is no effective control of this pest on asters.

Foot rots and wilts These diseases cause the plants to wilt and eventually die. Control by crop rotation and by improving the drainage of the soil.

CAMPANULA
Cutworms These soil-living caterpillars feed on the plant roots and on the base of the stems, often severing the latter. Control by raking in soil insecticides or by applying soil drenches of spray-strength general insecticide.

Froghopper These pests are easily spotted because of the frothy 'cuckoo spit' which covers the feeding larvae. Control with forceful sprays of general insecticide.

Red spider mite These tiny pests, which are only just visible to the naked eye, feed on the underside of the leaves causing the foliage to become discoloured. Control by repeat sprayings of derris, fenitrothion, formothion or pirimiphos-methyl.

Rust This disease shows up as orange-coloured spore pustules which develop on the underside of the leaves. Control by spraying with maneb, thiram or zineb.

CARNATION   *See* Dianthus.

CHEIRANTHUS (WALLFLOWER)
Cabbage root fly The grubs of this pest feed on the roots, stunting growth and causing the plants to wilt in hot weather. Control by dusting over the seed drills and around transplanted plants with bromophos, chlorpyrifos, diazinon, HCH or pirimiphos-methyl. Late attacks can be dealt with by the use of soil drenches of general insecticides.

Caterpillars These foliage-feeding pests can be controlled by the use of general insecticides.

Flea beetle These beetles bite small circular holes in the leaves and can be very destructive to young seedlings. Control by applying insecticidal dusts to the plants and to the surrounding soil.

Clubroot In plants infected with this disease, irregular swellings develop on the roots and the plants are stunted and liable to wilt in hot weather. Control by applying calomel (mercurous chloride) dust to the open drills and to the planting holes.

Downy mildew Infected leaves develop yellow blotches which show a greyish fungal growth on the underside. Control by repeat sprays of zineb.

CHRYSANTHEMUM
Aphids Greenfly and blackfly which attack chrysanthemums can be controlled by the use of greenfly killers or general insecticides.

Capsid bugs These insects feed on leaves, stems and buds, producing small brown calloused spots. Young leaves may be stunted and distorted and apical shoots may become 'blind'. Flowers, too, may be distorted. Control by repeat sprays of general insecticides in July and August.

Caterpillars Various species of caterpillar can cause serious damage by eating or skeletonizing the leaves. Control with general insecticides or with the biological insecticide, Bacillus thuringiensis.

Chrysanthemum blotch miner The yellowish maggots of this pest tunnel into the leaves forming a blister-like blotch. Control with repeat sprays of dimethoate, pirimiphos-methyl or trichlorphon.

Chrysanthemum eelworm This leaf-infesting eelworm produces large brown or black patches on the leaves. The best defence is to only use cuttings taken from eelworm-free stools. The pest can, however, be prevented from spreading to

Chrysanthemum blotch

the upper leaves by ringing the stem with vaseline early in the season and then periodically renewing this ring.

Chrysanthemum leaf miner The small grub of this pest tunnels into the leaf tissue producing whitish linear 'mines'. Attacks by this pest can go on all season but it can be controlled by repeat sprays of dimethoate, pirimiphos-methyl or trichlorphon.

Earwigs These pests damage both the foliage and the flowers. Control by trapping in inverted pots filled with straw or by spraying the plants and the surrounding soil with a general insecticide.

Froghoppers This pest is easily recognized by the frothy 'cuckoo spit' which covers the feeding larvae. Control with forceful sprays of general insecticide.

Leafhoppers These tiny, faintly coloured moth-like insects feed on the lower leaf surface causing a pin-point mottling. Control by spraying with a general insecticide.

Red spider mite These pests, which are only just visible to the naked eye, feed on the underside of the leaves, which become discoloured. It is not generally a major pest outdoors. Control by repeat sprays of derris, dimethoate, fenitrothion, formothion or pirimiphos-methyl.

Thrips These tiny insects feed on both the foliage and the flowers causing mottling of the leaves and a white flecking of the florets. Control by spraying with a general insecticide.

Slugs These pests feed on the foliage and stems but are easily controlled by the use of slug pellets.

Grey mould This disease flourishes in warm humid conditions when it can be highly damaging to the petals of both opening buds and open flowers. Initially the infection is limited to pin-point brown spots but it quickly spreads to give a general brown rot which is often covered by grey fungal growth. Control with benomyl, captan, dichlofluanid, thiophanate-methyl or thiram.

Foot and stem rot Plants infected with this disease become stunted and the lower leaves turn yellow and necrotic. In severe cases there is rotting of the roots and of the base of the stem. The best control is to grow the chrysanthemums on a different site although some control can be obtained with weekly soil drenches of zineb.

Petal blight This disease first shows as brown or pink spotting of the outer florets. These spots quickly turn to a brownish-black wet rot and join together to form larger decayed areas. In a short time the whole flower becomes infected and may then be invaded with grey mould. The damage is most serious in wet weather and in these conditions sprays of zineb should be applied, starting when the buds first show colour.

Powdery mildew This disease shows as a white powdery covering on the stem and leaves. Control by spraying with benomyl, dinocap, sulphur or triforine.

Ray blight This is mainly a greenhouse disease since it is most active in humid conditions and high temperatures. The infec-

80

Mildew on chrysanthemum leaf

tion first shows as a browning of the inner florets and then spreads to the whole flower. Cuttings may also be attacked, resulting in a rotting of the growing point or collapse of the stem. Protection against attack can be obtained by spraying with captan, maneb or triforine before the buds open or if the cuttings show signs of being attacked.

Rust This disease is not very common nowadays because modern varieties are resistant. Powdery brown spore pustules develop on the lower surface of the leaves with corresponding yellow spots on the upper surface. Control with sulphur, thiram, triforine or zineb.

White rust This disease can be recognized in the early stages by the presence of whitish-yellow, circular spore pustules on the underside of the leaves. Later these pustules turn brown and a light green depression appears on the opposite surface of the leaf. Suspected cases must be reported to the Ministry of Agriculture.

Virus Several types of virus attack chrysanthemums. The commonest and most serious is Aspermy Flower Distortion which causes a white flecking of the florets and may result in their distortion. Stunt virus, which leads to stunted growth and poor flower quality, is also common. The only approach to these troubles is to discard all infected stools.

CROCUS
Aphids The tulip bulb aphid overwinters under the dry outer scales of the corm and then moves on to the young shoot in spring. Control by dusting the corms with HCH prior to storage.

Bryobia mite This tiny pest, which is only just visible to the naked eye, feeds on the foliage causing it to become speckled, and later bronzed. Control by repeat sprays of derris, dimethoate, fenitrothion, formothion or pirimiphos-methyl.

Dry rot This disease causes the leaves to turn yellow and brown prematurely and, on close examination, it will be found that they are rotted at the base. Heavily infected corms also rot but those which are only slightly infected survive till the following spring when they grow and produce the characteristic symptoms. Control the spread of the disease by lifting and destroying infected corms.

Fusarium rot This common and serious disease is associated with premature yellowing and death of the foliage. It is caused by a soil-borne fungus which attacks the roots and the base of the corm, giving the latter a purplish colour. The covering scales also commonly turn olive-green. There are no effective counter-measures at present.

DAHLIA
Aphids Blackfly are serious and common pests of dahlias. Control by the use of greenfly killers or general insecticides.

Capsid bugs These bugs feed on the leaves, shoots and flower buds. Injured flower buds produce lopsided blooms whilst the stems are scarred and twisted. Damage to the young leaves

82

shows up as calloused brown spots at first. The leaves later become deformed and puckered. Control by spraying with a general insecticide.

Caterpillars Various types of caterpillar feed on the foliage and flowers. Control by spraying with a general insecticide or with the biological insecticide, Bacillus thuringiensis.

Earwigs These pests damage both the foliage and the flowers. Control by trapping the earwigs in inverted plant pots filled with straw, or by spraying with a general insecticide.

Grey mould This fungal disease can damage flowers and flower buds in humid weather. Control by spraying with captan or thiram.

Smut This disease first shows up as angular brown spots on the lower leaves and then spreads upwards, killing large areas of the leaf. Control by spraying with a copper fungicide.

DELPHINIUM
Chrysanthemum eelworm This leaf-infesting eelworm causes brown or black patches on the leaves. Some control can be obtained by applying vaseline rings to the stem early in the season and then periodically renewing them.

Delphinium moth Caterpillars of this moth damage the buds, leaves, flowers and seed pods. Control by spraying with general insecticides in June and early July when the eggs are hatching.

Leaf miner The small grubs of these pests tunnel into the leaves producing whitish 'mines'. Normally no control measures are needed against these pests but they can be controlled, if necessary, by the use of general insecticides.

Swift moth The soil-living caterpillars of this moth feed on the plant roots. Where this pest is common the plants should be lifted and the soil treated with a soil insecticide in the spring.

Slugs These pests are attracted to the young shoots so the use of slug pellets around the plants in spring is recommended.

Black blotch This bacterial disease shows in the form of large black blotches on the leaves, stems and flower buds. It is difficult to control but repeated sprays with a copper fungicide will protect young plants.

Powdery mildew Shows as a white powdery covering on leaves, stems and flowers. Control by repeat sprays of dinocap or copper.

DIANTHUS
D. barbatus (Sweet William)
Carnation fly The small grubs tunnel into the leaves forming faintly coloured 'mines'. Control by repeat applications of general insecticide, starting at the first sign of attack.

Leaf spot This disease first shows as faintly coloured circular spots on the leaves which are quickly killed. Control by repeat applications of a copper fungicide.

Rust Leaves infected with this disease develop circular rings of brown spore pustules. Control by repeat sprays of copper, thiram or zineb.

D. caryophyllus (Carnation)
Aphids Aphids can be troublesome on these plants but are readily controlled by the use of greenfly killers or general insecticides. The plants are also liable to attack by root aphids which stunt the growth and cause wilting in hot weather. These underground pests are more difficult to deal with but can be controlled by heavy soil drenches of spray-strength general insecticide.

Carnation fly The small grubs of this pest tunnel into the leaves, producing pale-coloured leaf 'mines' and eventually killing the leaves. They may also tunnel into the stems, causing the death of whole shoots. Control by repeat sprays of general insecticide, starting at the first sign of attack.

84

Apple powdery
mildew on leaves

Scab on Cox apple

A healthy apple and one
with brown rot

Cherry tree
with bacterial canker

Carnation thrips These tiny slender insects feed on the foliage and flowers causing a pale-yellow or silvery mottling of the leaves and a white flecking of the petals. Control with the use of general insecticides.

Ring spot This fungal disease shows as round grey spots on the stem and leaves which later develop concentric rings of spore pustules. It is difficult to control but repeated sprays with a copper fungicide are recommended.

Rust This disease can be recognized by the brown spore pustules which develop on both sides of the leaves and also on the stems. Control by repeat sprays of thiram or zineb.

ENDYMION
E. hispanicus (Bluebell)
Stem eelworm Infestation by this pest causes both the leaves and the flower stalks to become stunted. Pale streaks and patches also show up on the leaves. The only practical answer is to replant in a different area with clean healthy bulbs.

FRITILLARIA
Lily beetle Crown Imperial Fritillaria plants in Berkshire and Surrey can be severely damaged by this vivid red and black beetle since both the adults and the larvae feed on the leaves. Control by repeat sprays with a general insecticide.

GALANTHUS (SNOWDROP)
Aphids *See* Scilla.

Narcissus fly *See* Scilla.

Stem eelworm The leaves of infested plants show pale green streaks or patches together with local swellings (spickels). There is no easy method of control of this pest so the answer is to buy fresh stock and replant on a new site.

Grey mould The leaves and stalks are killed by grey mould, and black dots (sclerotia) develop on the bulb. Destroy infected bulbs and spray the remaining plants with captan.

GERANIUM *See* Pelargonium.

GLADIOLUS

Aphids Greenfly, which are common summer pests, are readily controlled by greenfly killers or by general insecticides. The corms may also become infested with the tulip bulb aphid which overwinters under the brown outer scales and then moves on to the developing shoot in spring. This pest can be controlled by dusting the corms with HCH.

Caterpillars These can be very damaging to the flower buds and flowers. Control by spraying with general insecticides.

Gladiolus thrips These tiny sucking insects cause yellow or silvery striping of the foliage and a white flecking of the petals. They overwinter on the corms so these should be dusted with HCH prior to storage. Summer attacks on the foliage and flowers can be dealt with by spraying with a general insecticide.

Swift moth The soil-living caterpillars of this moth attack the corms. Control by raking in a soil insecticide or by applying a soil drench of spray-strength general insecticide.

Core rot This fungus attacks the core of the corm and then spreads outwards as moist rot. As insurance against infection the corms should be thoroughly dried before being stored.

Dry rot This disease shows as black spots in the flesh of the corm. These spots later coalesce to give large, black areas and eventually the corm becomes mummified. The remedy is to plant only clean healthy corms and to practise crop rotation.

Hard rot and leaf spot Tiny brown or purplish-brown spots develop on the leaves and these quickly produce small, black fruiting bodies. As the spots enlarge, the centres turn grey whilst the outer rim remains brown. Infected corms develop water-soaked, dark-coloured spots. These increase in size and become sunken, well-defined lesions. Control by planting only clean healthy corms and by practising crop rotation.

Scab and neck rot This bacterial disease first shows on the leaves as minute reddish-brown specks. These enlarge to give circular or oval dark brown spots. Later the bases of the leaves may rot, causing the foliage to turn brown and wither. Spots also develop on the scales of the corm, whilst similar lesions appear on the base of the corms. Positive identification of this disease is given by the exudation of a yellow or brown gum from the corm lesions. Control by planting only clean healthy corms and by practising crop rotation.

HELLEBORE
Leaf spot Round or elliptical black blotches, marked with concentric zones, develop on the leaves. Later these run together into large patches and the leaves wither. The flowers may also be spotted. Control by removing diseased leaves and flowers and then applying repeat sprays of a copper fungicide.

HOLLYHOCK *See* Althaea.

HYACINTH
Aphids Infestations of greenfly stunt the plants and damage the flowers. Control with greenfly killers or general insecticides.

Narcissus fly The white grubs of this pest tunnel into the base of the bulb, destroying the flower bud and seriously damaging the rest of the bulb. Control by dusting the soil around the plants with HCH at fortnightly intervals from April onwards.

Stem eelworm The foliage of bulbs infested with these microscopic pests develops pale streaks and patches. Prevent attack by planting only clean healthy bulbs.

Grey bulb rot This disease infects the nose of the bulb causing a dry, grey rot and preventing emergence of the shoot. Prevent by buying only clean healthy bulbs.

IRIS
Aphids Leaf-feeding greenfly can be a problem but are easy

to control with greenfly killers or general insecticides. The bulbs may also become infested with the tulip bulb aphid which overwinters under the outer scales and then moves on to the young shoot in spring. This is controlled by dusting the bulbs with HCH prior to storage.

Caterpillars Various types of caterpillar feed on the foliage, eating holes in the leaves or chewing notches from the leaf margins. Control with general insecticides.

Narcissus fly The grubs of this pest tunnel into the bases of the bulbs destroying the flower buds and seriously damaging the rest of the bulb. Protect the bulbs from attack by dusting around the plants with HCH at fortnightly intervals from April onwards.

Leaf spot This disease attacks both bulbous and rhizomatous irises. It shows as brown oval spots on the leaves which later wither and die. Control by spraying with zineb.

LATHYRUS (SWEET PEA)
Aphids Greenflies are a major pest of these plants. Control with greenfly killers or with general insecticides.

Caterpillars These pests feed on the foliage, tying the young leaves together for protection. Spray with a general insecticide at the first sign of attack.

Bud drop The young flower buds turn yellow and drop off whilst still small. This trouble is not due to attack by pests or diseases but is caused by wrong cultural conditions such as a shortage of phosphate and potash or an excess of nitrogen. It is also favoured by waterlogged, badly aerated soil. It can be prevented by the addition of bulky organic materials, such as peat or composted bark, to the soil before planting and by the use of balanced fertilizers. It is also important to practise sensible watering.

LILIUM (LILY)
Aphids These pests not only cause damage by their feeding

but also transmit virus disease. Plants should therefore be sprayed with a greenfly killer or general insecticide at the first sign of attack. Other species which infest the bulbs should be controlled by dusting the bulbs with HCH prior to storage.

Lily beetle This vivid, scarlet and black beetle is mainly troublesome only in Surrey and Berkshire. It can be very destructive since both the adults and the larvae feed on the leaves, stems and seed pods. Control by repeat sprays with a general insecticide.

Thrips These minute slender insects feed on the leaves and flowers causing a yellow speckling of the leaves and a white flecking of the petals. Control with general insecticides. Lily thrips, which feed only on the bulb, can also be troublesome. These produce rust-coloured sunken spots on the base of the outer scales. Control by dusting the bulbs with HCH and storing them in cool conditions.

Leaf blight Oval-shaped, water-soaked spots develop on leaves infected with this disease whilst infected stems topple over. Control by spraying with a copper fungicide.

LUPINUS (LUPIN)
Caterpillars Stem-boring caterpillars sometimes attack lupins. These pests tunnel into the stems causing the shoots to collapse; the only method of control is to remove the damaged shoots and then to kill the caterpillar or pupa which is inside. Soil-living caterpillars, which feed on the roots, may also be a problem. These can be dealt with by the use of heavy soil drenches of spray-strength insecticide.

Rose thrips These tiny insects feed on the foliage causing a yellow flecking. Control with general insecticides.

Slugs These pests are attracted to the young shoots and need to be controlled by the use of slug pellets.

MARIGOLD
Aphids Blackfly can be very troublesome on marigolds. Control with greenfly killers or general insecticides.

Caterpillars Stem-boring caterpillars can kill whole shoots. Control by removing the damaged shoots and killing the caterpillars or pupae found inside.

Slugs These pests can be very damaging, especially to tagetes. Control by the use of slug pellets.

MATTHIOLA (STOCK)
Cabbage root fly The white grubs of this pest feed on the roots, stunting the plant growth and causing wilting in hot weather. Control by applying bromophos, chlorpyrifos, diazinon, HCH or pirimiphos-methyl to the soil around the young plants. Late attacks can be dealt with by applying soil drenches of general insecticide.

Caterpillars Foliage-feeding caterpillars can be a problem. Control with general insecticides.

Flea beetle These beetles bite small circular holes in the leaves and can be very destructive to seedlings. Control with insecticidal dusts applied to the plants and to the surrounding soil.

Clubroot Plants attacked by this disease develop irregular swellings on the roots and their growth is stunted. Control by dusting the open seed drills and the planting holes with calomel (mercurous chloride) dust.

Downy mildew The leaves develop yellow patches which show a greyish fungal growth on the underside. Control by repeat sprays of zineb.

MICHAELMAS DAISY
Caterpillars Root-feeding caterpillars can damage the plants. Control by heavy soil drenches of spray-strength general insecticide.

Cyclamen mite An attack by this pest in the early stages of growth results in stunted and deformed shoots, twisted leaves and a russeting of the stems and leaves. Flower buds are then

92

attacked and are either killed or produce only small, deformed flowers. Control by repeat sprays of derris, dimethoate, fenitrothion, formothion or pirimiphos-methyl.

Powdery mildew This disease shows as a white powdery coating on the leaves. Control by repeat sprays of dinocap.

MONKSHOOD *See* Aconitum.

NARCISSUS
Bulb scale mite These microscopic mites feed on the bulb scales causing brown streaks and making the bulb rather soft and light in weight. Plants from infested bulbs are lacking in vigour and die-back early.

Narcissus fly The white grubs of this pest burrow into the base of the bulb, destroying the flower bud and causing a wet rot. Infested bulbs either fail to grow or produce only thin, grassy leaves. Protect the bulbs from attack by dusting round the plants with HCH at fortnightly intervals from April until June.

Stem eelworm This microscopic pest infests both the bulb and the top growth. The foliage of infested plants is pale in colour, stunted and twisted, with small yellow swellings on the leaves. Flowers are produced late and die when the bulb finally rots. Infested bulbs feel soft and have dull-looking outer scales. When the bulbs are cut across, rings of brown decayed tissue can be seen. Buy sound, healthy new bulbs and replant on a fresh site.

Fire This disease can be identified by the presence of reddish-brown, elongated spots on the leaves and a brown spotting of the flowers. It is carried over in the bulb, and shoots from infected bulbs are diseased on emergence. These 'primary infectors' should be removed and the plants sprayed with thiram at 10-day intervals.

Leaf scorch This disease shows as a scorching of the leaf tips. Control by repeat sprays of zineb.

Smoulder This bulb-rotting fungal disease also causes a rot of the foliage and flowers in wet seasons. Control by spraying with zineb.

NYMPHAEA (WATER LILY)
Aphids The water lily aphid disfigures the foliage, distorts the stems and discolours the flowers. These pests can be controlled with greenfly killers or with general insecticides but these pesticides must not be used if there are fish in the pond.

Caddis fly The larvae, which are covered by a protective case of sand or vegetable debris, can be damaging to roots, buds, leaves and shoots. Ornamental pools should be drained and cleared of vegetable debris annually. Fish also help to keep the larvae under control.

Caterpillars The caterpillars of the brown china moth are unusual in that they can live in water. Control by hand picking.

Water lily beetle Both the adults and the larvae are very destructive to water lilies. One method of control is to hose the leaves, washing the insects into the water for the fish to feed on. Alternatively the leaves can be kept submerged by weights for a few days.

NASTURTIUM *See* Tropaeolum.

PAEONY
Leaf and bud eelworm This microscopic pest enters the leaf tissue causing a yellow or purple blotching. There is no practical control other than ringing the stems with vaseline to prevent the upward spread of the eelworms.

Swift moth Caterpillars of this pest gnaw into the tubers. Some control can be obtained by the use of heavy soil drenches of spray-strength general insecticide in May, June and July.

Wilt Some shoots turn brown at the base and then wilt.

94

Leaves of other shoots may also develop angular brown patches and the flower buds, too, may turn brown and die. Control by cutting off the wilted shoots below soil level. A copper dust should be applied to the crowns. Repeat sprays of dichlofluanid, starting shortly after emergence, also give good results. Similar repeat treatments with captan, thiram or zineb have given good results with tree paeonies.

PANSY *See* Viola.

PELARGONIUM (GERANIUM)
Capsid bugs These pests damage the leaves by their feeding punctures which develop into brown calloused spots. Attacked leaves later become puckered and torn. Control by repeat applications of general insecticide.

Caterpillars These pests can damage the foliage. Control with general insecticides.

Black leg This disease, which shows as a blackening of the base of cuttings, is caused by a soil-borne fungus. It can be prevented by the use of sterile growing compost.

Rust Infected leaves develop brown spore masses on the underside. These pustules are commonly arranged roughly in rings. Control by fortnightly applications of thiram or zineb.

PHLOX
Aphids Greenfly not only weaken annual phlox by their feeding but also stunt them and cause distortion of the flowers. Control with greenfly killers or general insecticides.

Caterpillars These foliage-feeding pests can be controlled by the use of general insecticides.

Chrysanthemum eelworm This microscopic pest enters the leaves causing large brown or black patches. The upward spread of the pest can be prevented by ringing the stems with vaseline but the best defence is to establish disease-free plantings on fresh soil.

Cutworms These soil-living caterpillars feed on the bases of the stems. Control by raking-in soil insecticides or by applying soil drenches of spray-strength general insecticide.

Froghopper These pests are easily spotted because of the frothy covering of 'cuckoo spit' surrounding the feeding larvae. Control by forceful sprays of general insecticide.

Stem eelworm These microscopic pests feed on the stems and leaves of perennial phlox. Infested plants can be recognized by their narrow, strap-like leaves. Their stems, too, are distorted and may show vertical cracks. Clean stock can be raised from root cuttings and the young plants established on a new site. Infested soil should be kept free of weeds and the site should not be planted with polyanthus, galanthus, hyacinths, narcissi, sallas or tulips.

Powdery mildew This disease of perennial phlox shows up as a white powdery covering on the leaves. Control by repeat sprays of dinocap.

PINKS *See* Dianthus caryophyllus.

POLYANTHUS
Bryobia mite These tiny pests, which are only just visible to the naked eye, feed on the foliage causing a yellow mottling of the leaves. Control by repeat sprays of derris, dimethoate, fenitrothion, formothion or pirimiphos-methyl.

Caterpillars These feed both on the foliage and on the flowers. Control with general insecticides.

Red spider mite Description and control measures as for 'Bryobia mite'.

Stem eelworm *See* Phlox.

Vine weevils The adult beetles feed at night on the foliage biting characteristic deep notches in the leaf margins. Control by spraying the plants and the surrounding soil with a general insecticide.

Leaf spots These fungal diseases show as various sized, discoloured spots. Control by repeat sprays of copper fungicides.

POLYGONATUM (SOLOMON'S SEAL)
Lily beetle This vivid red and black beetle is largely restricted to Surrey and Berkshire. It can be very destructive because both the adults and the larvae feed on the foliage. Control by repeated treatments with a general insecticide.

Solomon's seal sawfly In the south of England the caterpillars of this sawfly can cause serious damage. They eat elongated holes between the veins of the leaves and also from the margins. Control by spraying with a general insecticide at the first sign of attack, taking care to spray the underside of the leaves.

PRIMULA, GENERAL (*See also* Polyanthus)
Caterpillars These pests can seriously damage the leaves and flowers. Control with general insecticides.

Clay-coloured weevil This night-feeding beetle bites holes in the leaves. Control by spraying the plants and the surrounding soil with a general insecticide.

Root aphid These pests feed on the roots, weakening the plants and causing them to wilt in hot weather. Control by heavy soil drenches of a general insecticide applied at the normal spray-strength.

PYRETHRUM
Caterpillars These feed on the flowers, tying the florets together for protection. Control by hand picking or by the use of a general insecticide.

Chrysanthemum eelworm This microscopic eelworm enters the leaf tissue causing brown or black patches. The upward spread of the infestation can be prevented by ringing the stems with vaseline. The better answer, however, is to plant up new stock on a fresh site.

RANUNCULUS
Leaf miner The grubs tunnel into the leaves forming light-coloured mines but have little effect on plant growth.

Root-lesion eelworm These microscopic pests feed on the roots causing small wounds which act as entry points for bacteria and fungi which in turn rot the roots. Infested plants grow badly. Discard infested stock and re-plant on fresh soil.

RUDBECKIA
Caterpillars These foliage-feeding pests can be very damaging. Control with general insecticides.

Chrysanthemum eelworm *See* Pyrethrum.

Froghopper These pests are easily recognized by the frothy 'cuckoo spit' which covers the feeding larvae. Control by forceful sprays of general insecticide.

SALVIA
Capsid bugs Plants infested with these pests show brown calloused spots on the leaves. Later the leaves may become torn and ragged. Injury to the growing points leads to distorted growth or even to the shoots becoming 'blind'. Control with general insecticides.

SCILLA
Aphids The tulip bulb aphid overwinters on the bulb under the outer scales and then moves on to the developing shoot in spring, causing it to be stunted and deformed. Control by dusting the bulbs with HCH prior to storage.

Narcissus fly The grubs of this pest tunnel into the base of the bulb destroying the central tissue and flower bud. Control by dusting round the base of the plants at fortnightly intervals from April onward.

Stem eelworm Infested plants show pale streaks and patches on the foliage. There is no effective easy method of control so the best answer is to buy new bulbs and replant in a different area.

SNOWDROP *See* Galanthus.

SOLOMON'S SEAL *See* Polygonatum.

STOCK *See* Matthiola.

SWEET PEA *See* Lathyrus.

SWEET WILLIAM *See* Dianthus.

TROPAEOLUM
T. majus (Nasturtium)
Aphids Greenfly and blackfly are extremely common pests of this plant. Control with greenfly killers and with general insecticides.

Flea beetle These beetles bite small circular holes in the leaves and can be very damaging to seedlings. Control by applying insecticidal dusts to the plants and to the surrounding soil.

Tarnished plant bug These bugs feed by perforating the leaves, causing brown calloused spots. Damaged leaves may later tear and become ragged. Control with general insecticides.

TULIPA (TULIP)
Aphids Greenfly not only damage the plants by their feeding but also transmit the tulip break virus which causes the flowers to develop green or white streaking of the petals. Control with greenfly killers or with general insecticides. The bulbs may also become infected with the tulip bulb aphid which overwinters under the brown outer scales. These aphids move on to the young shoots in spring causing severe stunting and distortion. Control by dusting the bulbs with HCH.

Stem eelworm Damage by this microscopic pest shows up best at flowering when pale or purplish streaks develop on one side of the stem just below the flower. The infection then spreads to the petals where the affected parts remain green or pale. In severe attacks the stem splits and bends over. Infested bulbs are not easy to identify so the best policy is to buy quality bulbs from a reliable supplier.

Fire disease This is a common and serious disease which can be carried over both on the bulbs and in the soil. Young leaves are scorched brown at their tips whilst older leaves and petals show small spots. Remove heavily infected plants and give repeat sprays with dichlofluanid, thiram or zineb.

Grey bulb rot This soil-borne disease attacks the neck of the bulb causing a grey rot which prevents the emergence of the shoot and eventually kills the bulb. Buy new stocks and replant in a different area.

VIOLA
V. odorata (Sweet Violet)
Aphids Greenfly can be troublesome but are readily controlled by greenfly killers or general insecticides.

Leaf and bud eelworm Infestation by this microscopic pest stunts and deforms the growth. Dig up and destroy infested plants. Healthy new stock should then be planted on a new site.

Violet leaf midge The faintly coloured maggots of this pest feed on the leaves which roll inwards and become very swollen. These pests are difficult to control and the best approach is to pick off and destroy infested leaves.

Violet smut This disease produces large gall-like swellings on the stems and leaves and later these galls become filled with black spores. Control spread by removing and destroying the galls before they open.

V. ×Williamsii (Viola and Violetta) See V. × Wittrockiana.

V. × *Wittrockiana (Pansy)*

Aphids Small greenfly which feed on the underside of the leaves can be very damaging. The signs of attack are yellowing and wilting of the leaves. These pests are best controlled by a systemic greenfly killer, such as pirimicarb, since it is difficult to spray the underside of the leaves in their wilted condition.

Leaf spot This disease shows as small, circular, light-coloured spots on the leaves. Control by repeat sprays with a copper fungicide or with zineb.

Stem rot This disease causes a brown rotting of the stem base and can kill the plants. Plants grown in wet soil are most susceptible to attack, so bulky organic materials, such as peat or composted bark, should be worked into the soil before planting in order to improve drainage and aeration. Calomel (mercurous chloride) dust should also be sprinkled into the planting holes.

WALLFLOWER *See* Cheiranthus.

WATER LILY *See* Nymphaea.

ZINNIA

Capsid bugs These pests feed by puncturing the plant tissue causing brown calloused spots on the leaves, buds and flowers. Control with general insecticides.

Caterpillars Stem-boring caterpillars tunnel into the stems causing the death of the shoots. Remove and destroy damaged stems.

Grey mould This disease can cause a rot of both flowers and stems in wet weather. Control by repeated sprays of benomyl.

6 Lawns

MAJOR PESTS

Chafer beetles The grubs of these beetles feed on the grass roots and cause extensive damage. Lawns are further disfigured by birds tearing out the grass in search of the grubs. Control by soil drenches of spray-strength HCH applied in summer when the grubs are feeding near the surface.

Leatherjackets These grubs, which are the larvae of crane-flies (Daddy-long-legs), feed on the grass roots and on the base of the stems and can cause major damage to the turf. When fully grown, leatherjackets are 2·5–3·8cm (1–1½in) long but they are difficult to see in the soil because of their grey-black colour. They do most damage in the spring but can also be active in mild periods in the winter, and during wet weather in early summer. Since both the eggs and the larvae are sensitive to drought they are not usually troublesome in light soils and in hot, dry seasons. They are, however, common in the moister parts of the country and on badly drained lawns. Control by soil drenches of spray-strength HCH applied in humid weather in autumn and spring.

Earthworms Some, but by no means all, gardeners regard these as major pests of the lawn. Certainly the worm casts produced by some species can be a nuisance, particularly in spring and autumn. On the other hand, earthworms do help to aerate the soil and to improve the drainage by their burrowing. Some control of the worm population can be obtained by always collecting the clippings when mowing. Worms are also inhibited by the use of mosskillers based on ferrous sulphate,

Aphids on rosebuds

The effect of leaf-rolling sawfly on rose

Honey fungus

and by the application of top dressings of peat. One positive method of control is to apply materials such as carbaryl, derris, mowrah meal or potassium permanganate which bring the worms to the surface where they can be swept up and destroyed. Another approach is to apply chlordane which kills the worms below ground and has a long-term controlling action. Should, however, there be any risk of contaminating pools or water containing fish then the only safe method is to use potassium permanganate at 17g per 4·5 litres per sq m (½oz/gal/sq yd) and to sweep up the worms when they come to the surface.

OTHER PESTS
Ants Small ant-hills can be a problem, particularly on sandy soil. Control by dusting the individual hills with an antkiller based on chlordane, HCH or pirimiphos-methyl.

Dor beetle Small mounds of earth thrown up by this beetle when emerging from the soil can be seen occasionally. No control necessary.

Fly larvae The small larvae of fever flies and St. Mark's flies are sometimes found in clusters just below the surface of the turf. They can be mistaken for small leatherjackets but can be distinguished by their brown heads. The larvae themselves do little damage but the lawn may be disfigured by birds tearing up the grass in search of the larvae. Control with the treatments recommended for leatherjackets.

Mining bees These solitary bees can be a problem on sandy lawns because of the conical soil casts at the nest entrance. Infested lawns should be brushed prior to being mown.

FUNGAL DISEASES
Dollar spot This shows as small brown patches about 5cm (2in) across which later become bleached. This disease is most damaging to sea marsh turf, especially when the vigour of the grass is low. Control by feeding the lawn and applying a lawn fungicide such as benomyl, mercurous chloride or quintozene.

Fairy rings These show up as circular bands of darker, more vigorous grass. In some types there is only a single ring of stimulated grass but in the more serious cases there are two rings separated by a bare zone. Fairy rings are caused by the activities of soil-borne fungi which produce toadstools in late summer or early autumn. They can only be eliminated by sterilizing the underlying soil with formalin. First remove the turf over the stimulated zone or zones and then dig over the stripped area. Water the dug soil with formalin solution containing added wetting agent, using 4 litres of 40% formalin dissolved in 50 litres of water to treat 10sq m (6 pints/10gal/10sq yd). Then cover the treated area with polythene film and allow fumigation to go on for 7–10 days. Finally remove the cover and lightly fork over the ground and leave it exposed for a further 2 weeks before adding soil to restore the levels prior to re-seeding or re-turfing.

Fusarium patch This is the commonest and most widespread lawn disease. It is most damaging in September and October but can also develop during mild spells in winter and again in spring and early summer. The first sign of attack is the appearance of small yellow or brown patches about 5cm (2in) across. Later these may run together to give larger damaged patches. In mild, moist weather a pink or white cottony fungal growth develops on the affected grass. Control by applying repeat treatments of benomyl, quintozene or mercurous chloride, starting at the first sign of attack.

Ophiobolus patch This disease first shows as small bleached or bronzed patches. These areas gradually become larger over the years and the centres become filled with coarse grass. The soil-borne fungus responsible for this disease is favoured by alkaline soil and by wet conditions. Control with turf fungicides based on phenyl mercury acetate.

Red thread (Corticium) This fairly common disease attacks in late summer. It can be detected by the general pink colour which develops on the grass and by the appearance of coral-red fungal 'needles' on the infected leaves. Fortunately this disease is not lethal to the grass. It is best dealt with by a combination of feeding the lawn and the application of a lawn fungicide such as benomyl.

Seedling diseases Various soil-borne fungi can cause the grass seed to rot or the seedlings to collapse after emergence. Control by the use of seed dressings based on captan or thiram.

OTHER TROUBLES
Blue-green algae These appear as a bluish-green, gelatinous growth on the grass. These invasions are usually associated with water-logged turf which has been subjected to heavy rolling. Regular spiking to aerate the soil and to improve surface drainage is the best method of preventing this trouble.

Slime moulds These show as a whitish or yellowish jelly-like covering on the grass foliage. Later the jelly dries out to give minute, grey-coloured, sporing capsules. Slime moulds cause little damage and disappear when the turf dries out.

7 Greenhouse and Indoor Plants

Pest populations can build up with alarming rapidity in the greenhouse because continuous breeding is possible in this protected environment, and overwintering losses are reduced. The spread of disease is also favoured by the relatively high humidity and by the plants being crowded together. Indoor decorative plants are equally under constant threat of attack by pests although the drier atmosphere is generally less favourable to the spread of disease. In this situation the control of pests and diseases involves a combination of hygienic measures, good cultural practice and the sensible use of pesticides.

Each of the major pests and diseases has a wide range of host plants and these all show somewhat similar symptoms of attack. It is therefore more convenient to list the pathogens alphabetically rather than to deal separately with the individual host plants.

SOIL-BORNE PESTS
Eelworms Two types of these microscopic pests are involved. These are the cyst-forming and the root-knot eelworms. Plants infested with either type become stunted and show a tendency to wilt but it is necessary to examine the roots in order to confirm that eelworms are involved. Cyst-forming types produce white, yellow or brown pin-head sized cysts on the roots. These can be troublesome on cacti, carnations, Ficus species, stocks, wallflowers and tomatoes. It is normally recommended that infested plants should be destroyed but cacti can be saved by cutting off the roots, washing the base and then re-rooting in sterile compost.

The roots of plants infested with root-knot eelworms show galls ranging in size from pin-head to about 2.5cm (1in) in diameter. The range of plants attacked by these pests includes begonia, cacti, carnation, chrysanthemum, coleus, gloxinia and narcissus. Fortunately these pests do not survive drying so they are not normally transmitted on properly dried bulbs and corms. Other infested plants should be destroyed although leaf and stem cuttings can be taken if sterile compost is used.

Glasshouse symphylid These small, white, wingless insects are about 8mm ($\frac{1}{3}$in) long and have twelve pairs of legs and long antennae. They feed on plant roots and on any leaves which touch the ground, causing the plants to become stunted and to have a tendency to wilt. Symphylids are active creatures which quickly burrow into the soil if disturbed so they are not easy to spot. One way of checking for their presence is to dig up a suspect plant and then drop it into a bucket of water when any symphylids which are present will float to the top. One way of preventing damage by these pests is to mix diazinon into the soil before planting out. Alternatively, heavy soil drenches of spray-strength HCH or pirimiphos-methyl will keep them in check.

Root aphid These root-feeding pests can be very damaging to pot plants, causing them to become stunted and to wilt. Ideally the plant roots should be washed free of soil and then dipped in spray-strength general insecticide before being repotted. Alternatively, heavy soil drenches of insecticidal solution can be applied.

Sciarid fly (Fungus gnat) The small white grubs of these gnats feed both on dead organic matter and on plant roots. Established plants, when attacked by these pests, make poor growth and may even wilt, whilst seedlings and young rooted cuttings may be killed. Stocks of potting compost should be kept in sealed containers to prevent them becoming infested with the larvae. Potted plants which show signs of being infested should be given soil drenches with spray-strength HCH, malathion or pirimiphos-methyl.

Vine weevil The fat, whitish larvae, which are 6–10mm ($\frac{1}{4}$–$\frac{3}{8}$in) long, can cause serious root damage. They also tunnel into bulbs, corms and tubers. Plants which are liable to be damaged by vine weevils include azalea, begonia, cineraria, coleus, cyclamen, ferns, hydrangea and lilies. Control by the measures given under 'Sciarid fly'.

Woodlice Although these creatures normally feed on rotting vegetation they can damage plants by feeding on the roots and the base of the stems. Control by eliminating their hiding places and by spraying or dusting the soil with a general insecticide.

Slugs These night-feeding pests can be very destructive in the greenhouse, killing seedlings and young plants. Control by putting down slug pellets based on either methiocarb or metaldehyde.

OTHER PESTS
Aphids A wide variety of greenfly are liable to attack greenhouse and indoor plants. These pests are readily controlled with greenfly killers or with general insecticides.

Caterpillars Foliage-feeding caterpillars can be a problem on a wide range of greenhouse plants. Plants should be inspected regularly and treated with a general insecticide at the first sign of attack.

Glasshouse leafhopper These tiny, pale-yellow, winged insects have two V-shaped dark bands on the back. They attack many greenhouse plants including fuchsia, geranium, heliotrope, pelargonium, primula and verbena. Leafhoppers feed on the underside of the leaves, causing a yellow speckling of the upper leaf surface and checking the growth. Control by repeat sprays of general insecticides.

Glasshouse thrips These small slender insects attack a variety of plants including arum lily, azalea, chrysanthemum, citrus, ferns, fuchsia, orchids, palms and roses. Both the adult and young thrips feed on the underside of the leaves, on young

110

shoots and on flowers, causing a silvery speckling of the foliage and a white flecking of the flowers. The plants are further disfigured by the young thrips excreting drops of brown liquid which later become covered with a brown mould. Thrips are also important in the transmission of various virus diseases. Control by repeat applications of a general insecticide or by the use of HCH smokes.

Glasshouse whitefly These tiny, white, moth-like insects and their larvae feed on the underside of the leaves. Not only do they weaken the plants by their feeding but they also disfigure the foliage by producing quantities of sticky honeydew which later become covered with sooty moulds. Whitefly are difficult to control because the eggs and some of the larval stages are resistant to most insecticides. They are best controlled by a programme of three or more sprays of biores-methrin, pirimiphos-methyl or resmethrin applied at intervals of 3–4 days. Repeat treatments with HCH smokes are also recommended.

Leaf eelworms These microscopic pests feed inside leaf tissue, producing brown or black patches between the veins. Various species attack a variety of plants including chrysanthemum, begonia, ferns, lily and primula. Violets are also attacked but in this case the damage shows as a crinkling and distortion of the young leaves and flowers. Infested plants should be destroyed and pots, boxes and staging should be sterilized. The upward spread of the pest on some valuable plants can be prevented by ringing the stems with vaseline.

Leaf miners The tiny grubs of these pests tunnel into the leaf tissue, producing faintly coloured mines or blotches. They can be troublesome on azalea, carnation, chrysanthemum and polyanthus. Control by repeat sprays with general insecticides.

Mealy bugs These bugs, which can be recognized by their white, powdery wax covering, are widespread pests on a large variety of greenhouse plants. They not only weaken the plants by their feeding but also produce large quantities of sticky honeydew which later become covered with sooty moulds. Foliage-feeding mealy bugs can be controlled by repeat sprays of general insecticides. Some species, however, are root-feeders and for these the control measures are as for root aphids.

Red spiders and other mites These pests, which are only just visible to the naked eye, feed on the foliage of a wide variety of plants. Infested leaves become discoloured, distorted and may even be scarred. Bud drop may occur on impatiens. Control by repeat sprays of derris, dimethoate, fenitrothion, formothion, malathion or pirimiphos-methyl.

Scale insects These pests, which are fixed firmly to the plant surfaces, can be recognized by their shell-like scaley covering. Like all sucking insects, they produce disfiguring honeydew which becomes covered with sooty moulds. Scale insects on woody tissue can be removed by scraping, whilst foliar infestations can be controlled by repeat sprays of malathion or pirimiphos-methyl.

SOIL-BORNE DISEASES
Foot rots These soil-borne diseases cause a rotting of the base of the stem and lead to the death of the plant. Foot rots attack many plants including aster, begonia, dahlia, dieffenbachia, peperomia, poinsettia, cucumber and tomato. Since there is no really effective cure for this trouble, the infected plants should be dug up and destroyed. Further attacks can be prevented by soil sterilization and by the use of sterile growing media.

Root rots and wilts These diseases can affect a variety of decorative plants such as azalea, carnation, chrysanthemum and dahlia, as well as food crops such as cucumbers and tomatoes. The first sign of attack is a greying or yellowing of the foliage. This is then followed by wilting and the death of the plant. In the absence of any effective curative treatment the only approach to this problem is to grow the plants in sterilized soil or in sterile growing compost.

Freesia leaf spot

Seedling diseases Soil-borne fungi can kill germinating seeds and can also cause the damping-off of seedlings. The risk of damage to germinating seeds can be reduced by the use of captan-based seed dressings. Thin sowing, coupled with sensible watering, reduces the incidence of damping-off but, should this trouble develop, then the best approach is to water the soil with Cheshunt compound.

OTHER DISEASES
Grey mould (Botrytis) This fungal disease causes a soft rot of stems, leaves and flowers. The affected parts, which soon become covered with a grey mould, later dry out and turn grey-brown. The spread of this disease can be checked by increasing the ventilation and by avoiding late night watering of the plants. Protect against attack by spraying with benomyl, captan, thiophanate-methyl or thiram.

Leaf spots Many greenhouse plants are subject to attack by leaf-spotting fungi. Control by picking off infected leaves and then applying protective sprays of general fungicide.

Powdery mildews These diseases show up as a white powdery covering on the stems, leaves and flowers. Susceptible plants include antirrhinum, begonia, carnation, chrysanthemum and hydrangea. Control by repeat sprays of benomyl, dinocap, triforine or sulphur.

Virus diseases These produce a wide variety of symptoms including stunted and deformed growth, variations in leaf form and colouring and colour breaking of the flowers. Since it is not possible to control viruses by the use of chemicals, the only answer is to destroy infected plants and to replace them by clean healthy stock. It is also essential to control insects such as aphids and thrips which spread viruses. This is most important since some apparently healthy plants can carry viruses which are extremely damaging to other species of plant.

GREENHOUSE HYGIENE
Good housekeeping is an essential factor in the maintenance of plant health in the greenhouse. The floor and benches should be kept clear of debris and all dead and damaged plant parts should be removed and burned. This reduces the spread of disease as also does the thorough washing of plant pots and seed trays prior to their being stored away for future use.

Overwintering pests pose a special problem and these can only be dealt with effectively by temporarily emptying the greenhouse. The glass and internal structure should then be washed down with disinfectant solution and afterwards the house should be fumigated by burning sulphur candles. These produce toxic sulphur dioxide gas which effectively sterilizes the whole interior.

Soil in the borders needs to be sterilized annually with formalin to prevent the build up of root rots, stem rots and wilts. This, however, is a lengthy and laborious process and the easier approach is to grow the plants in pots, growing bags or in ring culture rather than directly in the border soil.

Appendix– Directory of Chemicals

Insecticides

Common chemical name	Product name
Bacillus thuringiensis	Herbon Thuricide HP
bioresmethrin	Combat Whitefly Insecticide
bromophos	PBI Bromophos
carbaryl	Boots Garden Insect Powder
chlordane	Chlordane 25
	Murphy Chlordane Wormkiller
	Nippon Ant Powder
chlorpyrifos	Murphy Soil Pest Killer
derris	Abol Derris Dust
	PBI Liquid Derris
diazinon	Combat Soil Insecticide
dicofol/dinocap/maneb/ pyrethrum	Murphy Combined Pest and Disease Spray
dimethoate	Boots Systemic Greenfly Killer
	Murphy Systemic Insecticide
DNOC/petroleum oil	Ovamort Special
fenitrothion	PBI Fenitrothion
formothion	Topguard Systemic Liquid
HCH (BHC)	Boots Ant Destroyer
	Fumite Smoke Cone
	Lindex Garden Spray
HCH/captan	Murphy Combined Seed Dressing
HCH/tecnazene	Fumite Tecnalin Smoke
malathion	Murphy Liquid Malathion
menazon/HCH	Abol-X
nicotine	XLAL Insecticide
oxydemeton-methyl	Greenfly Gun
pirimicarb	{ Rapid Greenfly Killer
	{ Rapid Aerosol

pirimiphos-methyl	ICI Antkiller
	Sybol 2
	Sybol 2 Dust
pirimiphos-methyl/	Sybol 2 Aerosol
pyrethrins	Kerispray
pyrethrum	Anti-Ant Powder
resmethrin/pyrethrum	Bio Sprayday
sodium tetraborate	Panant
tar oil	Mortegg
trichlorphon	Dipterex

Slug Killers

metaldehyde	Boots Slug Destroyer
	ICI Slug Pellets
methiocarb	Draza
	Slug Gard

Fungicides

benomyl	Benlate
bupirimate/triforine	Nimrod-T
captan	Orthocide Captan Fungicide
captan/HCH	Murphy Combined Seed Dressing
carbendazim/maneb	Combat Rose Fungicide
chloraniformethan	Milfaron
copper fungicides	Cheshunt Compound
	Murphy Liquid Copper Fungicide
dichlofluanid	Elvaron
dinocap	Murphy Dinocap Smoke
	Toprose Mildew Spray
folpet/dinocap	Murphy Rose Fungicide
formaldehyde	Murphy Formaldehyde Soil Steriliser
mercurous chloride	ICI Club Root Control
	PBI Calomel Dust
sulphur	Lime Sulphur
technazene	Fumite Technazene Smoke
thiophanate-methyl	Murphy Systemic Fungicide
thiram	ICI General Garden Fungicide
zineb	Dithane

116

Index